Real Science-4-Kids

SCIENCE

Book 2

Rebecca Woodbury, Ph.D., M.Ed.

RS4K

RS4K Real Science-4-Kids

Illustrations: Janet Moneymaker

The Building Blocks Series

Science Book 2 Student Textbook (hardcover)
ISBN 978-1-941181-80-5

Published by Gravitas Publications Inc.
Real Science-4-Kids®
www.gravitaspublications.com

Table of Contents

Physics

Geology

CHAPTER 1

The Tools of SCIENCE

1.1 Introduction

How do scientists study the world around them? How do they know how fast an electron travels or what the back of a butterfly wing looks like? How do they see faraway planets or measure the size of an earthquake?

Scientists use **basic** and **advanced tools (instruments)** to measure, weigh, calculate, and explore the world around them. An advanced scientific tool, or **instrument**, is a tool that is more complicated than a basic tool and is used to do detailed and exact work. Basic tools and instruments help scientists crack open rocks, magnify small animals, and measure air temperatures.

Tools are a very important part of doing science. Without basic and advanced tools, scientists would not be able to cure diseases and create new materials, and they would not have been able to invent computers.

1.2 Brief History

Both basic and advanced tools began as inventions to help people control and understand their surroundings. An **invention** is something new that is created to help people do a certain task.

For example, if there were no ladders, you might find it hard to get apples from the top of a tree. This might lead you to think about different ways you could get the apples. If you rolled a nearby boulder over to the tree, you might be able stand on it to reach the branches. Or you might think of using ropes to pull yourself up. By thinking about ways to reach the apples and experimenting with your ideas, you are "inventing" ways solve the problem.

The same thing happens when scientists invent tools to help them study the world around them. They think of tools that might allow them to see details better, hear sounds better, or sense temperatures better. Then they create the tools that will help them make these kinds of observations.

Tool making has a very long history.

Many historians think the first tools that people invented were those used in controlling water so food could be grown. The ancient land of Mesopotamia, located between the Tigris and Euphrates rivers, was one area where tools were used to bring water to farms. (This area is now the country of Iraq.) Early farmers in Mesopotamia used walls and channels to guide water into fields to grow food. The early Egyptians and Chinese used dams, canals, and artificial lakes to control water for growing food.

Because ancient people were always testing new ways to make their living conditions better, the tools they invented were continually improved upon and adapted. With these new tools, ancient people were able to explore better ways to grow food, how to build taller buildings, and how to use metals for weapons. Ancient people also developed tools for making beautiful sculptures, glass objects, pottery, and wooden furniture.

It is hard to pinpoint exactly when and where the first instruments were used, but instruments also began in simple forms that were improved on over time. The advanced scientific instruments of today make it possible for scientists to understand and explore more and more of the world around us. Advanced scientific instruments allow scientists to explore very small things, such as atoms, and also very big things, such as the Sun.

1.3 Basic Tools

The most important tools scientists use are their own **five senses: sight**, **hearing**, **smell**, **touch**, and **taste**.

To explore the world around them, scientists make many observations by using just their five senses. For example, scientists can observe the size of an elephant and the color of a frog.

Scientists can observe that a car travels faster than a snail and that stars shine in the night sky.

However, there is a limit to how much information the five senses can provide. Basic tools **extend** the senses. By using basic tools, scientists can make objects appear larger, measure liquids more accurately, and determine the temperature of a body of water. Some **basic tools** include **magnifying lenses**, **thermometers**, **beakers**, and **hammers**.

With these basic tools, scientists can explore temperature, how ants carry food to their homes, and what minerals are inside a rock or geode. Basic tools are easy to use, and many are light and small enough to carry in a backpack.

1.4 Advanced Tools

Basic tools can only go so far in providing information to scientists. A magnifying glass is great for observing ants, but it won't work to see the cells inside the ant. A thermometer is great for measuring the temperature of a pond, but it can't measure the temperature of a faraway star. A hammer is perfect for opening up a rock and exposing the crystals, but it can't open up the crystals to expose the atoms.

In order to observe cells, measure the temperature of stars, or see atoms, more **advanced tools** are needed. An **instrument** is an advanced tool that allows scientists to do things such as taking more precise measurements or collecting more detailed data.

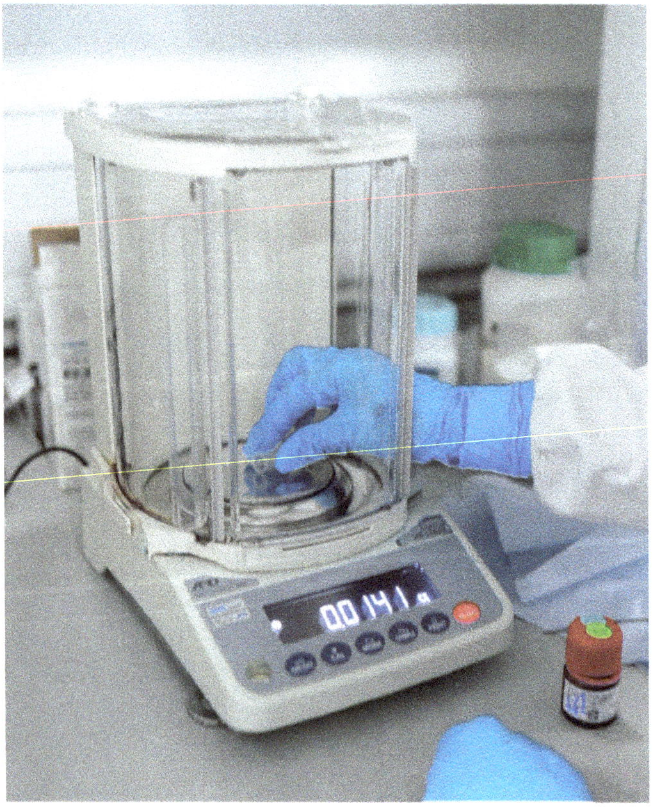

For example, as a scientist you might want to know exactly how much a tiny sample of salt weighs. To measure the weight, you would use a **scale** of some sort. In general, a basic scale won't work with a tiny amount, so you would need to use a more advanced **analytical scale**. Maybe you are doing an experiment with electricity and need to find out how many electrons are traveling through wires. To do this you would need to use a **voltmeter** to measure the flow of electrons. Or maybe you want to see how many atoms are on the surface of a piece of carbon. You would use a **scanning tunneling microscope** to see the atoms you want to count.

Gold atoms
seen with scanning tunneling microscope

These and many other instruments help scientists measure, weigh, and observe features that can't be observed with basic tools or their own five senses.

1.5 Summary

- **Basic and advanced tools** help scientists explore the world around them.
- Ancient people used **tools** to control water and grow food.
- **Basic tools** extend the **five senses** — **sight**, **hearing**, **smell**, **touch**, and **taste**.
- **Advanced scientific tools (instruments)** help scientists observe features that **basic tools** cannot.

1.6 Vocabulary

Basic Words

advanced (uhd-VAANST)

voltmeter (VOHLT-mee-tuhr)

Challenge Words

analytical (aan-uh-LIH-tih-kuhl)

instrument (IHN-struh-muhnt)

invention (ihn-VEHN-shuhn)

magnify (MAAG-nuh-fiy)

thermometer (thuhr-MAH-muh-tuhr)

tunneling microscope (TUH-nuh-lihng MIY-kruh-skohp)

CHAPTER 2

Chemist's TOOLBOX

2.1 Introduction

How do you study tiny atoms? How do you see small molecules? How can you measure, weigh, and monitor chemical reactions?

Today, chemists use a variety of tools to study atoms, molecules, and chemical reactions.

The tools chemists use include special types of glass, metals, and equipment needed to study the smallest building blocks of matter. In this chapter we will take a look at some of these tools.

2.2 Brief History

The earliest chemists were called **alchemists**.

Alchemists tried very hard to turn metals, such as lead, into gold. They never got it to work, but they did learn about chemical reactions. Alchemists were not true chemists as we know them today, because alchemists did not use a scientific method. But alchemists learned a great deal about how different chemicals react with other chemicals.

Early chemistry tools and equipment used by alchemists were made of glass, clay, stone, and a variety of metals. One early chemistry tool called the **alembic** was used by early Greek and Arabic alchemists to make perfumes and medicines. Another early chemistry tool called the **pelican**

was similar to the alembic and was also used to purify materials for medicines, perfumes, and alcohol. Today, alembics and pelicans continue to be important tools in chemistry labs.

2.3 Basic Chemistry Tools

Chemists use many different kinds of tools and equipment. Most chemistry labs have both **basic tools** and **advanced tools (instruments)**.

▶ Chemists often need to measure the quantity of a **liquid**.

A liquid is any pourable substance, such as water. You can't measure liquid water with a ruler, so you have to put it into a container that will help you measure how much you have.

For example, when you bake a cake, you use a measuring cup to measure liquids such as water, milk, and vegetable oil. A chemist uses a basic tool called a **beaker** in the same way as you use a measuring cup. Like a measuring cup, a beaker has lines on it showing how much liquid

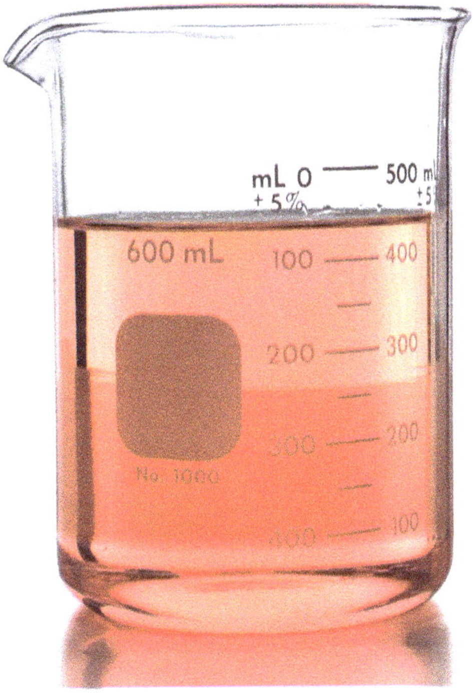

it contains, and it has a small spout for pouring. Beakers don't have a handle like a measuring cup, and the sides of beakers are straight up and down rather than being angled. In a chemistry lab, a beaker does the same job as a measuring cup.

- Simple **scales** and **balances** are basic tools found in chemistry labs.

To measure how much "stuff" there is, scales and balances are used. To find out how much of you there is, you would use a scale. When you go to the doctor's office, they have you step on a platform. They adjust some weights to show how much you weigh, or they read a number from a display. A scale is used to weigh you, just like a chemist uses a scale to weigh objects in the lab.

2.4 Advanced Chemistry Tools

All over the world there are many different types of chemistry labs studying different kinds of chemistry. Often you can find **advanced tools (instruments)** in these labs.

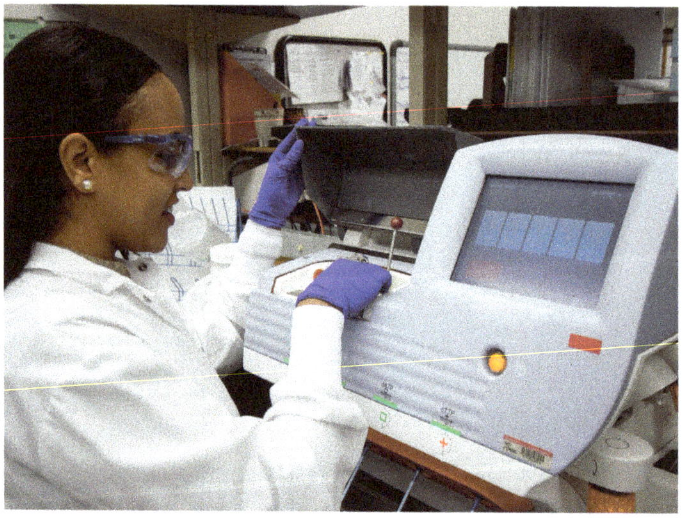

- For example, a lab that is studying gases might use a **gas chromatograph**.

A gas chromatograph is an instrument that can detect different types of gases in even very small amounts.

A gas chromatograph has long tubes connected to a computer or a machine that draws graphs. The chemist puts a gas in one of the tubes, the gas chromatograph tests it, and the results are displayed on a computer or are printed out.

Many chemistry labs research ways to make new chemicals. Some of their experiments are quite complicated. For example, sometimes a chemistry experiment might require that little or no oxygen gas come in contact with the chemicals, so specialized equipment is used to keep the oxygen out. These experiments require advanced **glassware** of different sizes and shapes. Like the alembic and pelican, some of this glassware has funny names, such as an **Erlenmeyer bulb** or a **Buchner flask**.

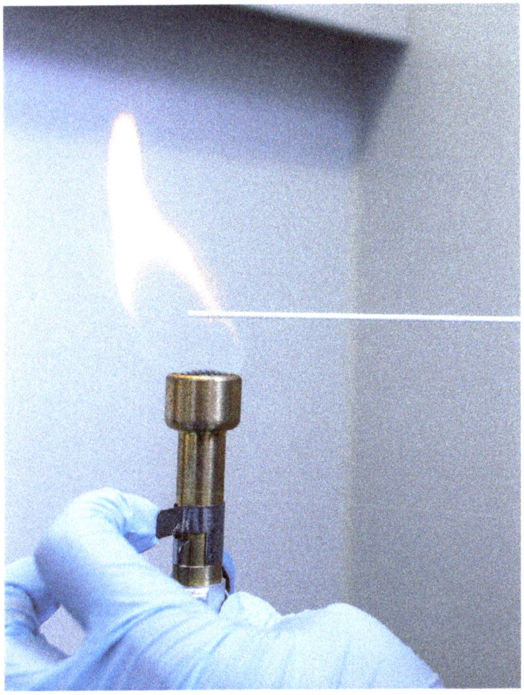

It is common for chemists to make their own **specialized tubes and glassware** to use in specific experiments.

In school, many young chemists learn how to blow, cut, and mold glass so they can create unique glassware for special experiments.

2.5 Summary

- Chemists use a variety of **basic tools** and **specialized advanced instruments** to do experiments.
- The early **alchemists** tried to turn lead into gold.
- Some **basic chemistry tools** are **beakers**, **scales**, and **balances**.
- Chemists sometimes make their own **specialized glassware** to use in experiments.

2.6 Vocabulary

Basic Words

balance (BAA-luhns)

beaker (BEE-kuhr)

scale (SKAYL)

Challenge Words

alchemist (AAL-kuh-mihst)

alembic (uh-LEHM-bihk)

Buchner (BUHK-nuhr)

chromatograph (kroh-MAA-tuh-graaf)

Erlenmeyer (UHR-luhn-miy-uhr)

pelican (PEH-lih-kuhn)

CHAPTER 3

Acids and BASES

3.1 Introduction

When **atoms** and **molecules** meet, they can trade places, join together, or separate from each other. A **chemical reaction** happens when atoms or molecules meet and any of these changes take place. A chemical reaction is one of the events chemists measure with tools.

In this chapter we will look at some special kinds of chemical reactions. These reactions are called **acid-base reactions**. Acid-base reactions are easy for chemists to study with some basic tools.

3.2 Acids and Bases Are Different

Have you ever noticed that when you bite a lemon it tastes sour and makes your cheeks pucker?

Have you ever tasted mineral water or baking soda water? They are not sour like a lemon. They are bitter or salty.

Have you ever noticed that soap is very slippery in your fingers, but lemon juice and vinegar are not?

The molecules in a lemon are different from the molecules in baking soda water or mineral water.

- Lemons have molecules in them called **acids**. It is the acid in lemons that gives them their sour taste.

- Baking soda water and soap contain molecules that are called **bases**. Bases often make things feel slippery or taste bitter.

3.3 H and OH Groups

Acids and bases are different from each other in many ways. This is because a base is a different kind of molecule than an acid. Acids and bases are different because they have different **atom groups**.

- A **base** has an **OH group** [say "O" "H" group]. An OH group is just an oxygen atom and a hydrogen atom together.

- Most common **acids** have an **H group** [say "H" group]. An H group is just a hydrogen atom.

We can see in the next picture that **sodium hydroxide** (a base) has an OH group and **hydrogen chloride** (an acid) has an H group.

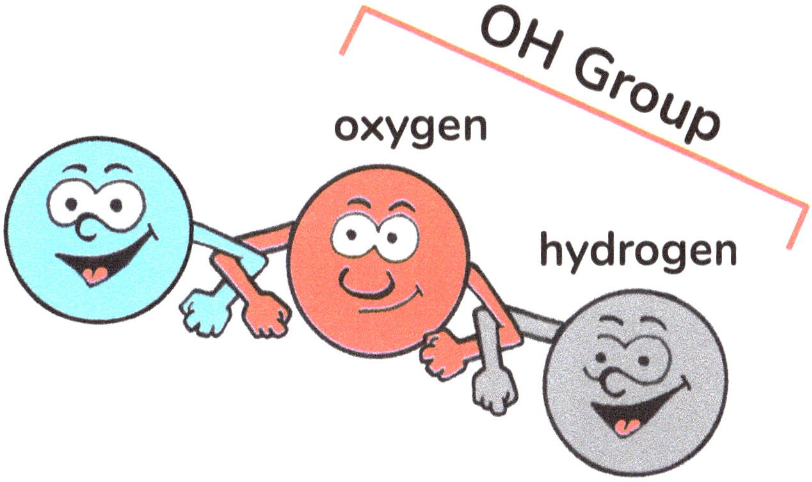

SODIUM HYDROXIDE
(a **base**)

HYDROGEN CHLORIDE
(an acid)

3.4 Both Are Important

Both acids and bases are very important. They are needed in lots of very useful chemical reactions. You have a strong acid inside your stomach to break down your food. Without the acid in our stomach, we could not digest our food.

- **Acids** are also found in batteries, lemons, oranges, grapes, and even soda pop.

All these things have **acids** in them.

Bases are found in lots of cleaners, like window cleaner, bathroom cleaner, and soap. They are also found in some foods like bananas and dates. Bases are even used to make your stomach feel better. We'll see why in the next chapter.

All these things have **bases** in them.

3.5 Observing Acids and Bases

Chemists can use basic tools to see an acid-base reaction. Some acids and bases give off heat or explode when they react. Other times we cannot tell when an acid-base reaction happens. When we can't see an acid-base reaction, we can put something into the acid-base mixture that will show us that the reaction is taking place.

- This "something" is called an **indicator** because it indicates, or tells us, something is happening or has happened.

We use different kinds of indicators all the time. Stop lights indicate when we can go or when we should stop. When we turn on the oven, an indicator tells us when it is hot enough. A thermometer is an indicator. It can tell when your body has a fever. Indicators are also used in chemistry.

INDICATORS

- An **acid-base indicator** tells us whether we have an acid or a base.

There are different kinds of acid-base indicators. A simple acid-base indicator is red cabbage juice. Red cabbage juice turns pink with acids and green with bases.

Red Cabbage Indicator

3.6 Summary

- **Acids** taste sour.
- **Bases** taste bitter and are slippery.
- **Acids** have an **H group,** and **bases** have an **OH group**.
- **Acids** and **bases** are found everywhere — in batteries, in your stomach, in household cleaners, and even in bananas and lemons!
- An acid-base indicator tells us whether we have an acid or a base.

3.7 Vocabulary

Basic Words

acid (AA-suhd)

base (BAYSS)

Challenge Words

hydrogen chloride (HIY-druh-juhn KLAW-riyd)

indicator (IHN-duh-kay-tuhr)

sodium hydroxide (SOH-dee-uhm hiy-DRAHK-siyd)

CHAPTER 4

Acids and Bases REACT

4.1 When Acids and Bases Meet

In the last chapter we learned about two different kinds of molecules — **acids** and **bases**. We saw that acids and bases are found in lots of different things. Acids are in batteries, lemons, and soda pop. Bases are in soap, window cleaner, and bananas too.

What happens when acids and bases meet? When molecules meet, sometimes they react. Do acids and bases react when they meet?

1 An acid and a base meet.

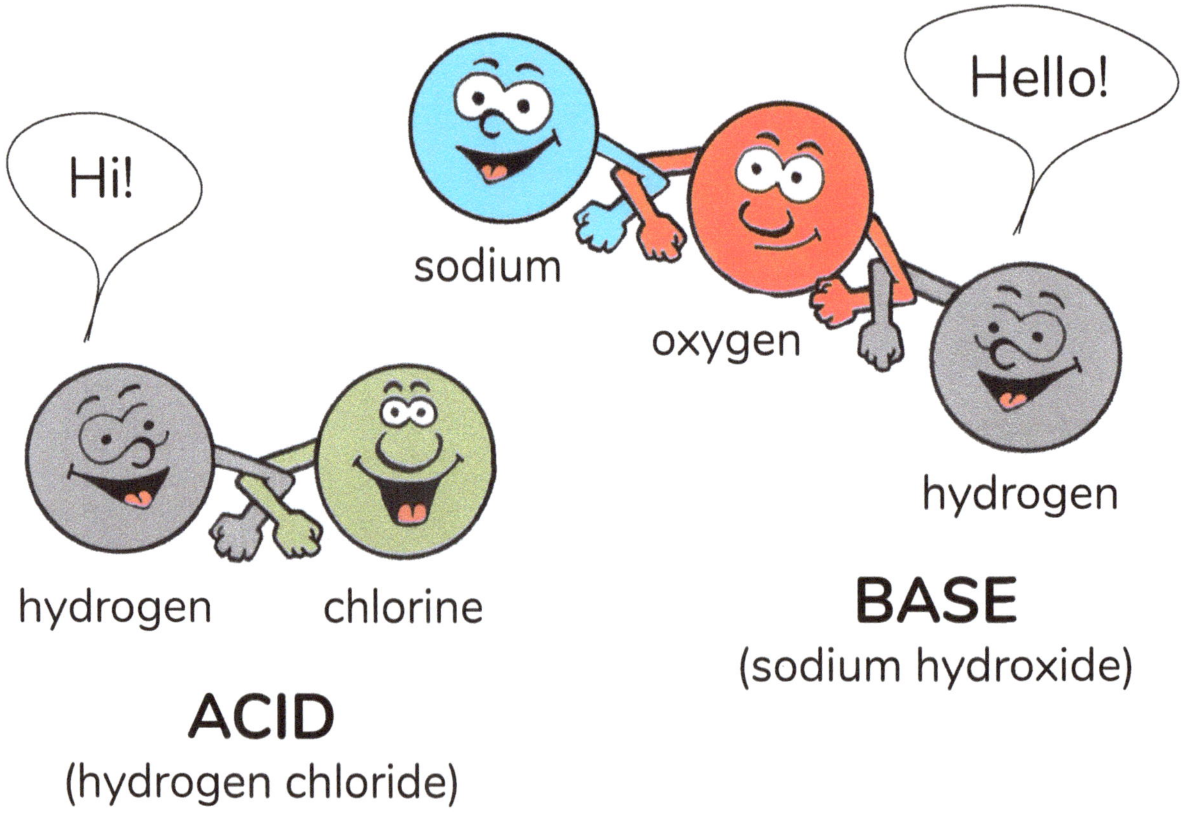

4.2 Acid-Base Reactions

In fact, they do! Acids and bases make a special kind of chemical reaction called an **acid-base reaction**. When an acid and a base meet, the atoms in the acid exchange with the atoms in the base.

After they meet, some atoms leave their molecules.

2 Atoms leave their molecules.

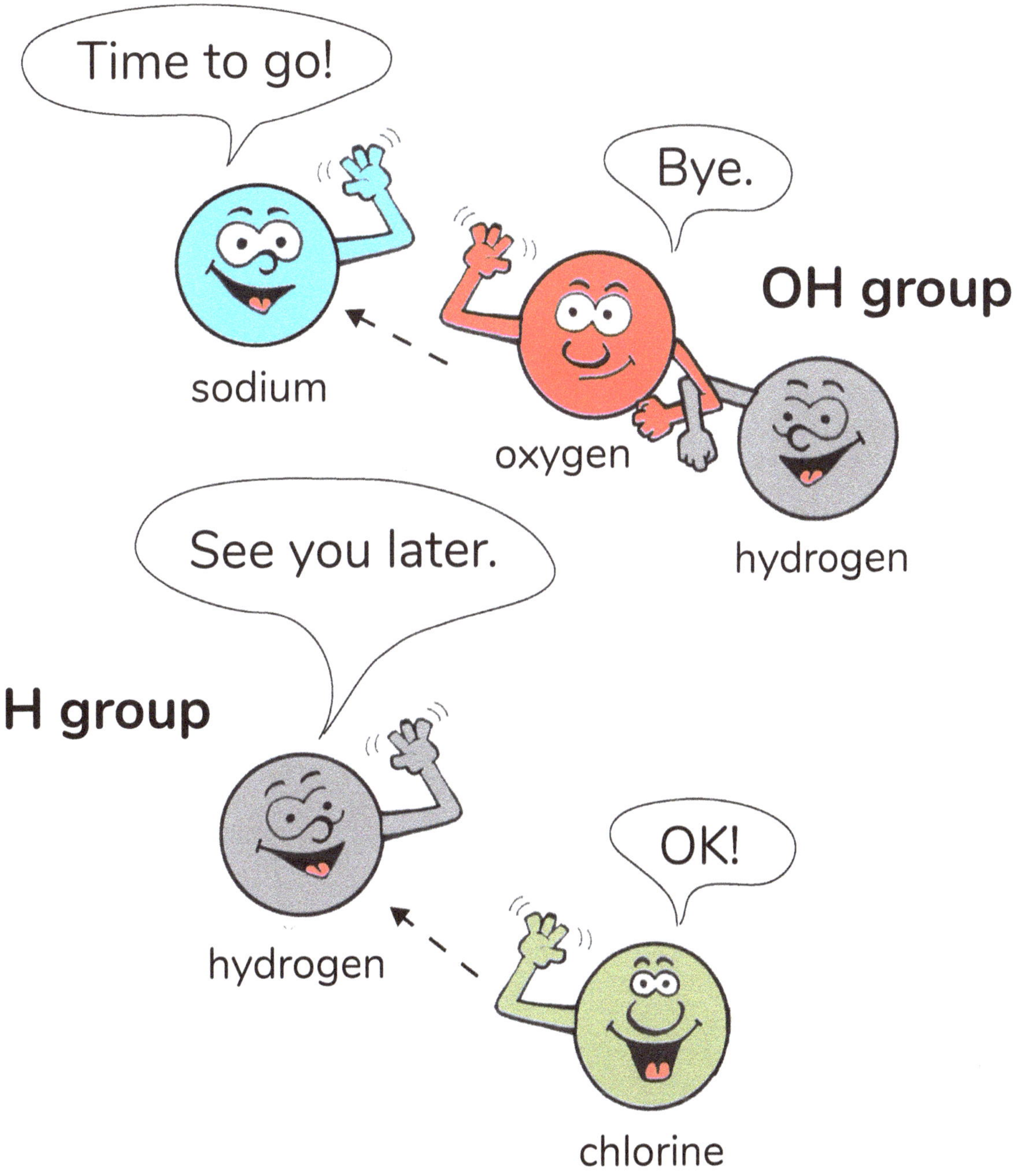

Next, the atoms that left their molecules go to the other molecules and "make new friends."

3 The atoms make new friends.

Now two new molecules have been made. The new molecules for this reaction are **water** and **table salt (sodium chloride).**

4 Two new molecules are made.

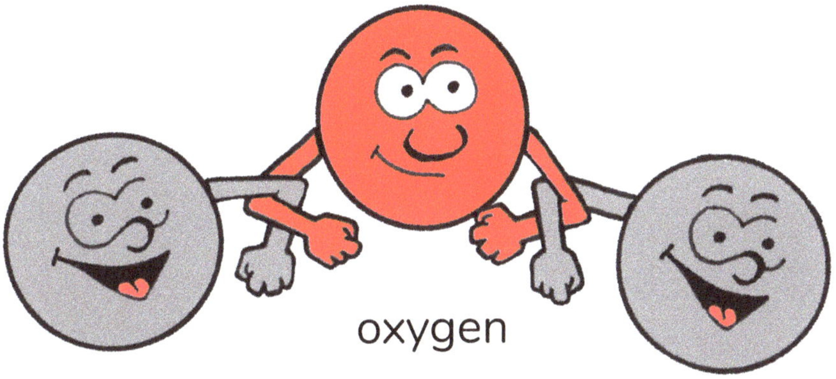

oxygen

hydrogen

hydrogen

Water

sodium

chlorine

Table Salt

(sodium chloride)

The acids and bases are no longer acids and bases. When they react, they become other kinds of molecules, such as salt and water.

4.3 Important Acid-Base Reactions

Acid-base reactions are very important. For example, your stomach has acid in it. This acid is necessary for digesting your food. Sometimes there is too much acid. When this happens, your stomach hurts. The medicine your mom or dad may give you is a base. It reacts with the acid in your stomach, turning it into a salt and water. That makes your stomach stop hurting.

4.4 Summary

- Acids and bases react with each other in **acid-base reactions**.
- When an acid and a base meet in an **acid-base reaction**, atoms in the acid exchange with atoms in the base.
- Many **acid-base reactions** make salts and water.
- **Acid-base reactions** are very important.

4.5 Vocabulary

Review Words

acid (AA-suhd)

base (BAYSS)

hydrogen chloride (HIY-druh-juhn KLAW-riyd)

reaction (ree-AAK-shuhn)

sodium hydroxide (SOH-dee-uhm hiy-DRAHK-siyd)

Basic Words

H Group (AYCH GROOP)

OH Group (OH AYCH GROOP)

CHAPTER 5

Food and TASTE

5.1 Tasty Molecules

Now you know why vinegar and lemons taste sour — they're acids. And why mineral water and soda water taste bitter — they're bases. Have you ever wondered why salt tastes salty and sugar tastes sweet?

We have learned that everything around us is made of atoms and atoms combine to make molecules. The food we eat is made of molecules, but not all of the food we eat tastes the same. Why? Different molecules in different foods make foods taste different.

1. **Acid molecules found in soda pop make soda pop taste sour.**

(Soda pop also has lots and lots of sugar in it, so it also tastes very sweet.)

Phosphoric acid

(molecule found in soda pop)

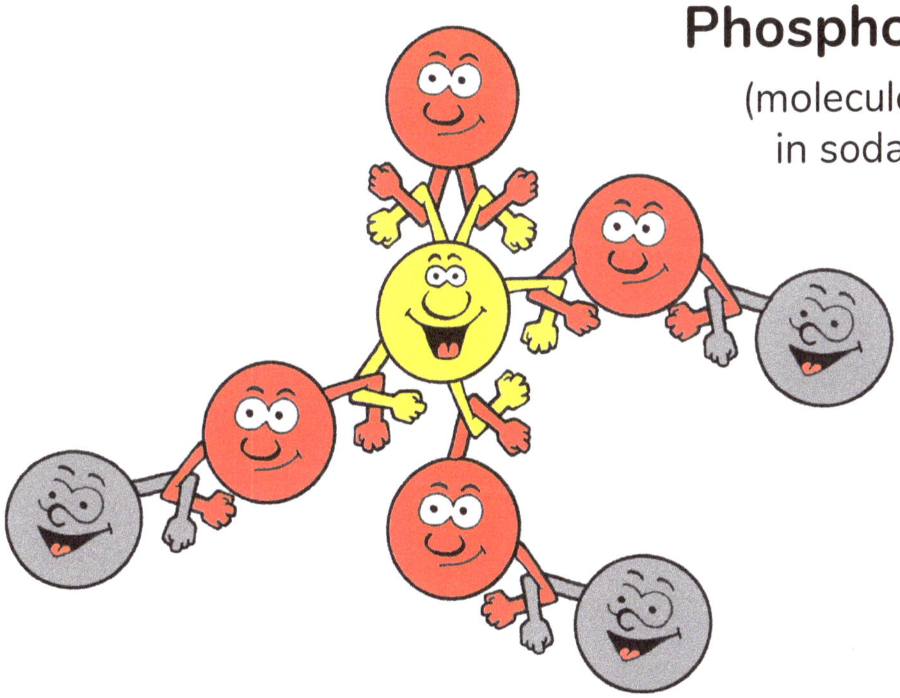

② Acid molecules in vinegar make vinegar taste sour.

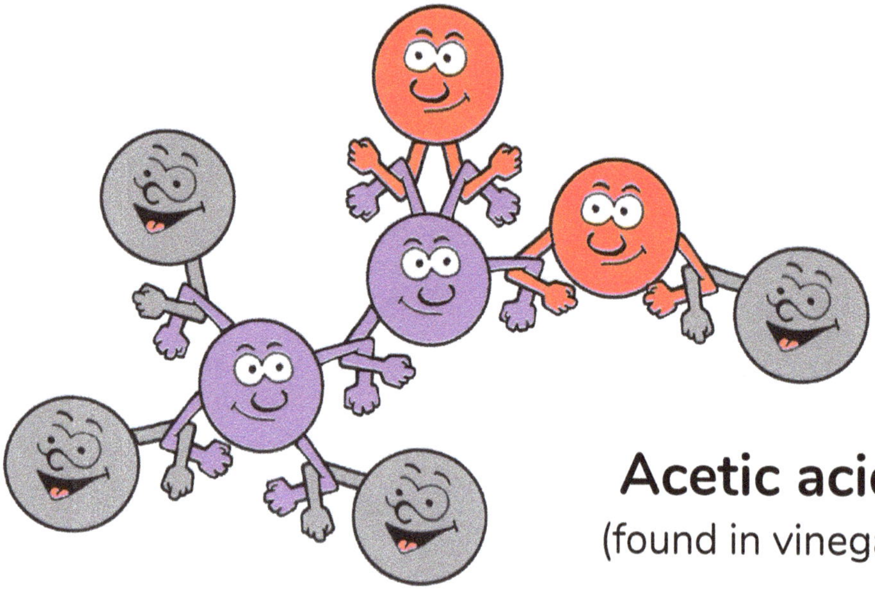

Acetic acid
(found in vinegar)

We have already seen that sour foods often have an acid in them. Lemons, vinegar, and grapefruit have acid in them. When you eat foods with acid in them, your tongue tells your brain "sour."

Foods that have salt in them taste salty. Salt molecules look very different from acid molecules. Remember that table salt has a sodium atom and a chlorine atom hooked together. When you eat foods with salt in them, your tongue tells your brain "salty."

Table Salt
(sodium chloride)

5.2 The Amazing Tongue

Your **tongue** is designed to tell your brain what kind of molecules are in your food. It can sense acids, bases, salt, sugar, and many other molecules. The tongue is essentially a tool your body uses to detect certain atoms and molecules.

Glucose
(a sugar)

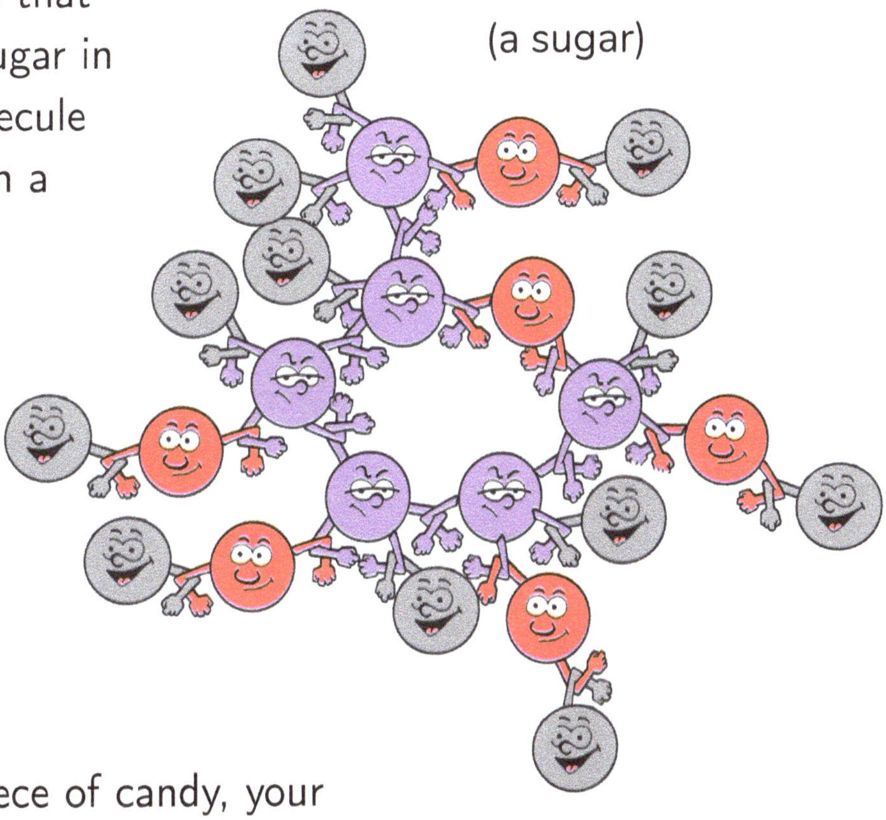

For example, foods that taste sweet have sugar in them. A sugar molecule looks different from a salt molecule or an acid molecule. A sugar molecule is larger, has more atoms in it, and some of the atoms are hooked together in a ring. When you eat a piece of candy, your tongue tells your brain "sweet" because your tongue senses sugar.

Your tongue can tell the difference between a salt molecule, an acid molecule, and a sugar molecule.

Your tongue is a remarkable **indicator.**

Remember from Chapter 3 that an acid-base indicator tells the difference between acids and bases. There are lots of man-made indicators, like acid-base indicators, thermometers, and stop lights, but there are no indicators as intricately designed as your own tongue.

5.3 Large Tasty Molecules

Even though your tongue is an amazing indicator, your tongue can't taste all molecules. For example, a raw potato doesn't exactly taste sweet, salty, bitter, or sour, but something in between. In fact, a potato is made mostly of sugar molecules, but your tongue can't taste the sugar molecules in a raw potato because the molecules are hooked together in long chains called **carbohydrates.**

Carbohydrate

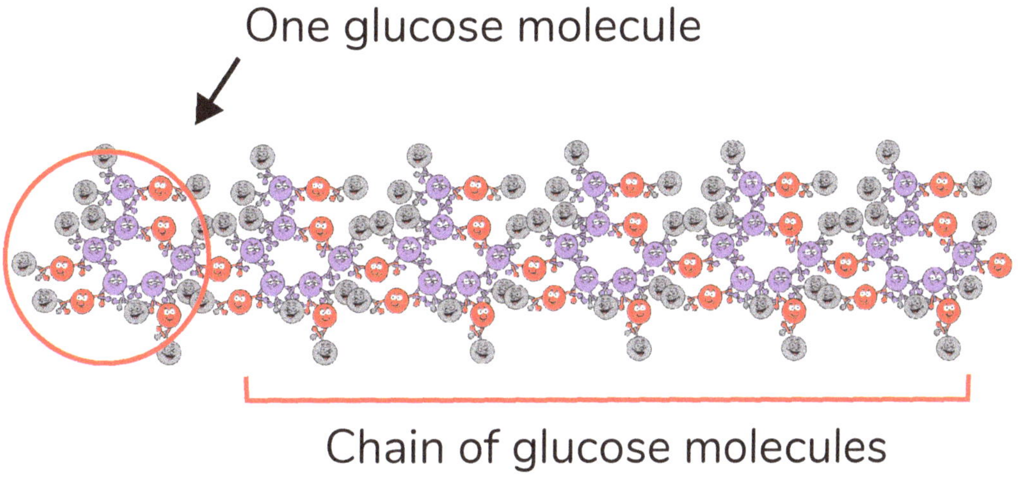

▶ Many different foods have carbohydrates.

Bread, rice, pasta, potatoes, and many fruits have carbohydrates in them. Carbohydrates are important molecules for your body. Because carbohydrates are made of sugar molecules, they provide the energy your body needs to ride a bike or climb a tree.

5.4 Summary

- Foods taste different because foods are made of different molecules.
- Your tongue is an amazing **indicator** that can tell the difference between salts, acids, bases, and sugars.
- **Carbohydrates** are long chains of sugar molecules.

5.5 Vocabulary

Basic Words

glucose (GLOO-kohs)

Challenge Words

acetic (uh-SEE-tihk)

carbohydrate (kahr-boh-HIY-drayt)

phosphoric (fahs-FAWR-ihk)

CHAPTER 6

Biologist's TOOLBOX

6.1 Introduction

How do you study living things? How can you look at cells or protists if they are too small to see when using only your eyes? How do you track a tiger when it can run fast and hide in the jungle? How can you measure, weigh, and monitor a hippopotamus when it is too large to fit on a scale?

Biologists use a variety of **tools** to **measure**, **monitor**, **weigh**, **track**, and **observe** living things. In this chapter we will take a look at some of the tools biologists use.

6.2 Brief History

People have been observing plants and animals for a very long time. Ancient people used both plants and animals for food, clothing, and shelter. Ancient civilizations learned about plants and how to create farms to grow food for large numbers of people. These civilizations also learned about animals and how to use them for transportation, how to raise them for food, and how to tame them for pets.

Although many people observed plants and animals, historians often call **Aristotle** the first real biologist. Aristotle was a philosopher who lived in Greece from 384-322 B.C.E.

ARISTOTLE 384-322 B.C.E.

Aristotle was the first person to organize plants and animals into categories. He was particularly interested in studying the *differences* between various plants and animals. He believed that the first step to understanding plants and animals is to look at all the ways in which they are different. He observed how animals with hair are different from animals without hair and how animals with paws are different from animals with hooves. His most valuable tool in studying differences was his ability to make **good observations**.

6.3 Basic Biology Tools

Today a biologist's most basic and important tool is the same as it was for Aristotle — the ability to make **good observations**. Biologists need to observe how plants and animals grow, live, and interact with other plants and animals. Many **field biologists** (those who work outdoors) take years to carefully observe and record things such as animal behavior, how different plants grow, and how changes in their surroundings affect the way plants and animals live.

- Another basic tool used by biologists is the **field notebook**.

A field notebook is essential for recording observations. Field notebooks can be large or small, print or digital. In the field notebook a biologist records the date, time of day, location, and the plants or animals being observed. Some biologists draw pictures directly in their field notebook, and many biologists carry a camera or video recorder to capture their observations.

A **magnifying glass** is a basic tool that can be very helpful for a biologist.

A magnifying glass makes small things appear larger. For example, if you leave your pie outside, you might discover in the morning that ants have come to have a taste.
You can see the ants with your eyes, but if you use a magnifying glass, you can see them even more clearly. They look big in the magnifying glass, and you can see the hair on their legs, the smoothness of their backs, and the number of pie crumbs they are taking.

6.4 Advanced Biology Tools

Biology covers a broad range of specialized subjects. For this reason there are many different kinds of **advanced tools** that biologists use.

Some biologists study plants and might use **kits** to test soils or water quality. Other biologists study animals and might use sophisticated **tracking equipment**, **cameras**, and **airplanes**. Some biologists who study animals that live in the oceans will use **boats**, **shark cages**, or **submarines**.

Many biologists use **microscopes** to see very small objects and tiny creatures. A microscope is a special kind of tool that helps you look at things too small for your eyes alone to see.

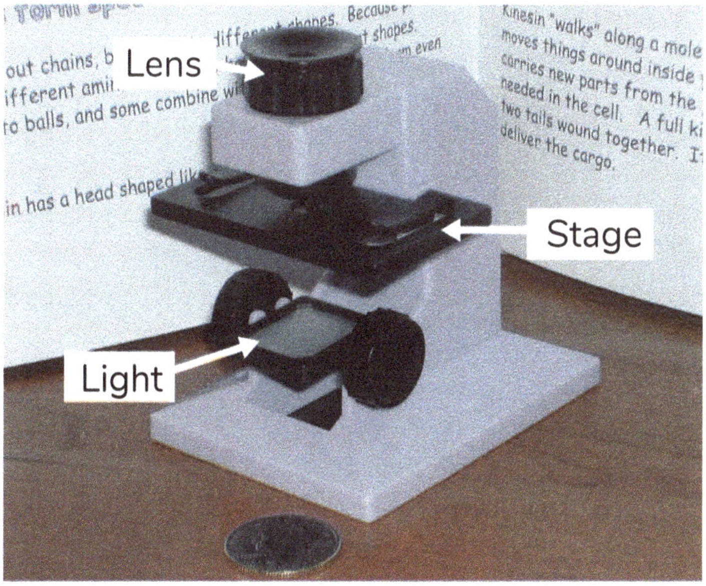

A **microscope** is like a magnifying glass in that it makes very small things look big. A microscope has a **lens**, usually made of glass. The lens **magnifies** (makes whatever you are seeing look bigger) like a magnifying glass does. But a microscope is also different from a magnifying glass. For example, a microscope has a **stage** where you put the **sample** (the thing you want to see—like pond water or ocean water). The stage helps keep the sample still while it is being observed.

Also, a microscope usually has a little light that makes the sample brighter and easier to see. In some microscopes, the light comes from a little mirror that captures room light and reflects it back through the microscope. Other microscopes use a tiny light bulb under the stage to illuminate the sample being observed.

6.5 Summary

- Biologists use a variety of **basic and advanced tools** to do experiments and make **good observations**.
- **Aristotle** was the first biologist who tried to organize plants and animals into groups.
- The most important **tool** for a biologist is the ability to make **good observations**.
- Biologists use lots of different **tools** to study plants and animals, such as **microscopes**, **boats**, **tracking equipment**, and **cameras**.

6.6 Vocabulary

Basic Words

observe (uhb-SUHRV)

weigh (WAY)

Challenge Words

Aristotle (EHR-uh-stah-tuhl)

biology (biy-AH-luh-jee)

magnify (MAAG-nuh-fiy)

microscope (MIY-kruh-skohp)

submarine (SUHB-muh-reen)

CHAPTER 7

Protists MOVE

7.1 Tiny Creatures

How tiny is the smallest living creature? If you look at yourself, you might think you are pretty small compared to your dad, but are you the smallest living creature? No! You are not as small as your cat. The cat is smaller than you are. But is a cat the smallest living thing? No. In fact, the cat chases (and sometimes eats) moths or birds or mice that are smaller than the cat.

How about a moth? Do you think the moth is the smallest living creature?

No. A ladybug is smaller than a moth, and an ant is smaller than a ladybug, and a gnat is even smaller than an ant. So how small is the smallest living creature?

- The smallest living creatures are so small that you can't even see them when looking with just your eyes. There is a whole world of tiny creatures that you can't see that live in ponds, in oceans, in dirt, and even inside of you!

- One type of tiny creature you can't see with only your eyes is called a **protist** (also called a **protozoan**).

A protist is a small creature that can do many of the things bigger creatures (like you) can do. Protists can crawl and swim and eat and sense light. Protists are very small but can do amazing things.

To observe protozoa (protists), you have to use a microscope. The first person to see protists was a man from Holland named **Anton van Leeuwenhoek**. With his microscope he saw little animals in pond water. He also found them in his mouth!

7.2 Different Kinds of Tiny Creatures

There are many different kinds of protists that can be seen only with a microscope. But how do you know what kinds of creatures you are seeing?

- Scientists sort living things into groups to make them easier to study.

The **kingdom Protista** is a very large group with over sixty thousand different kinds of protists. Because it is such a large group, scientists need to sort the protists into even smaller groups to study them.

7.3 Sorting Protists

One way to sort protists into smaller groups is to notice the different ways they move and then group them by how they move.

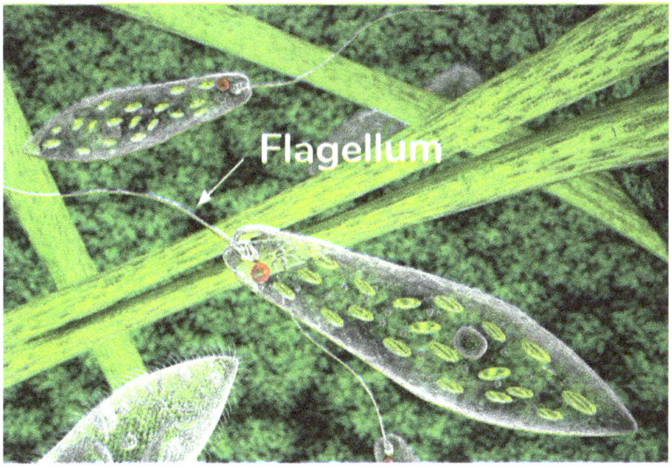

Euglena

One way protists move is to swim with a long tail called a **flagellum**. Some protists, like **euglena**, have a flagellum. Protists that move by using a flagellum are called **flagellates**.

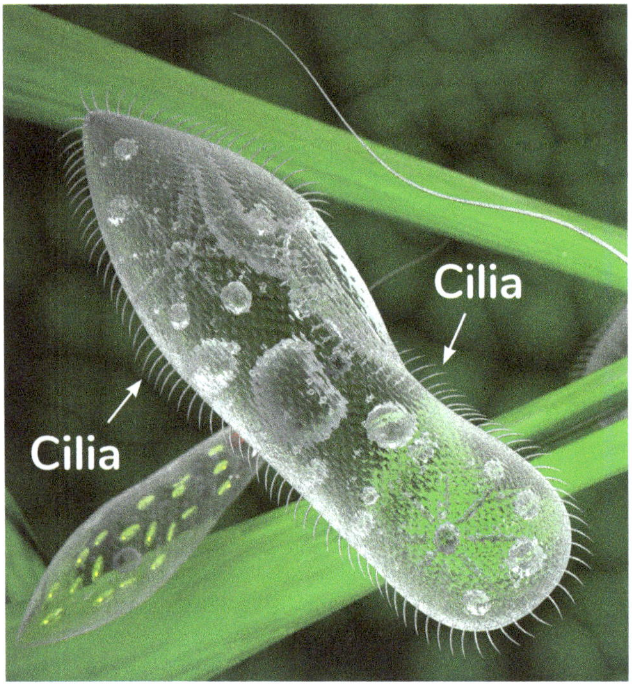

Paramecium

Another way protists move is by using small hairlike features called **cilia**. Cilia beat fast in the water, making the protists move forwards and backwards and sideways. A **paramecium** uses cilia to move. Protozoa that move with cilia are called **ciliates**.

Some protists move by crawling. **Amoebas** are protists that use false feet called **pseudopods** to crawl from one place to another. Protists that use false feet to move are called **amoeboids**.

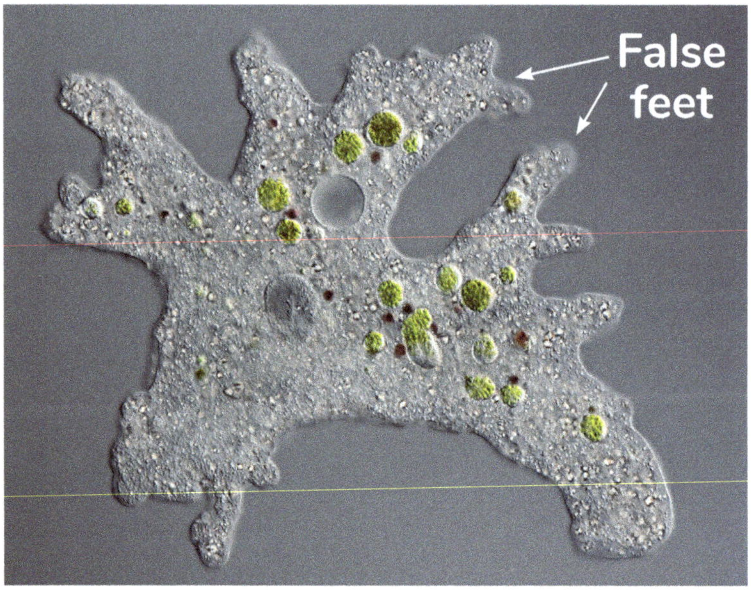

Amoeba

7.4 Summary

- There are many small creatures that we cannot see by using just our eyes.
- Scientists use a **microscope** to see small creatures.
- **Protists** (also called **protozoa**) are small creatures that can be found in pond water and ocean water.
- One way scientists sort **protists** is by how they move. Three groups of **protists** are **flagellates**, **ciliates**, and **amoeboids**.
- Many **protists** move by using a **flagellum**, by using **cilia**, or by using **pseudopods** (false feet) to crawl.

7.5 Vocabulary

Basic Words

protist (PROH-tihst)

Challenge Words

amoeba (uh-MEE-buh)

amoeboid (uh-MEE-boid)

Anton van Leeuwenhoek
(AAN-tahn VAAN LAY-vuhn-huuk)

cilia (SIH-lee-uh)

ciliate (SIH-lee-uht)

euglena (yoo-GLEE-nuh)

flagellate (FLAA-juh-luht)

flagellum (fluh-JEH-luhm)

paramecium (pehr-uh-MEE-shee-uhm)

Protista (proh-TIHS-tuh)

protozoan (proh-tuh-ZOH-uhn)

pseudopod (SOO-duh-pahd)

CHAPTER 8

Protists EAT

8.1 Euglena Eat

A euglena uses sunlight to make its own food. It changes sunlight to food by using **chloroplasts**. Chloroplasts are special parts inside a cell. A chloroplast contains **chlorophyll** which is a green substance that captures sunlight. Because chlorophyll is green, it gives euglena their green color.

A euglena will swim toward the sunlight that it uses to make food. A euglena has a little **eyespot** that helps it know where to find the sunlight.

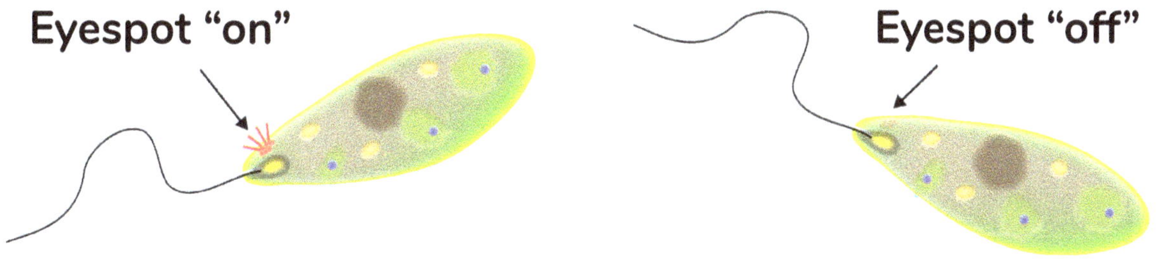

Can you tell which euglena is swimming toward the Sun?
How can you tell?

8.2 Paramecia Eat

Not all protists can make their own food like a euglena does. A paramecium has to go find its food, just like we do. But a paramecium cannot go to the grocery store for eggs and milk like we can. It must use its cilia to swim around and look for food in the water.

A paramecium eats other small creatures, such as other protozoa or bacteria. The paramecium has a mouth that it uses to capture food. The mouth does not move like a human mouth and it doesn't have any teeth.

The cilia around the mouth move, or beat, rapidly. This makes the water near the mouth of the paramecium swirl. Take your hands and move them in some water, and you can feel the water swirling around your hands.

A paramecium uses the swirling water to move food toward its mouth.

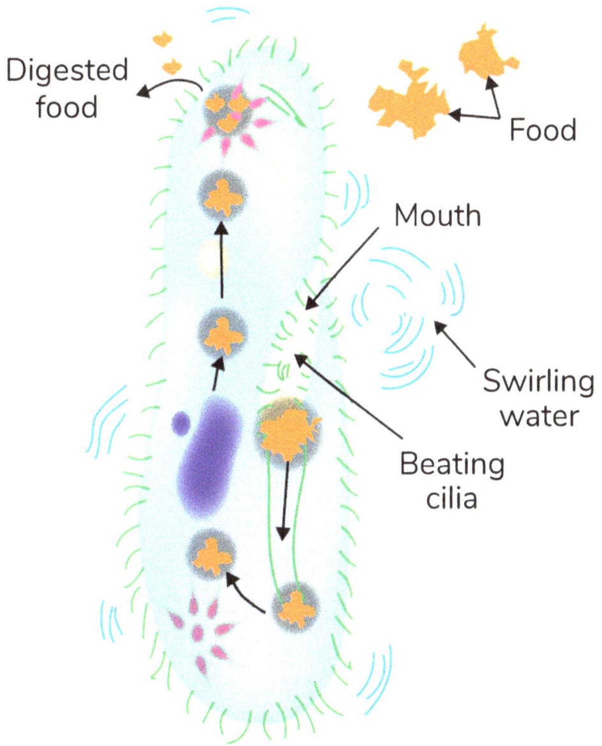

When the food enters the mouth of the paramecium, it travels through a small tube into a tiny stomach and gets digested. The paramecium takes what it needs from the digested food, and the unused food is pushed out through a small hole.

Although a paramecium has only one cell, it can move, eat, and digest food just like larger creatures. For being so small, the paramecium is an amazing creature.

8.3 Amoebas Eat

An amoeba eats with its feet! Can you imagine eating with your feet? It would be pretty hard for you to eat that way, but it isn't hard at all for an amoeba.

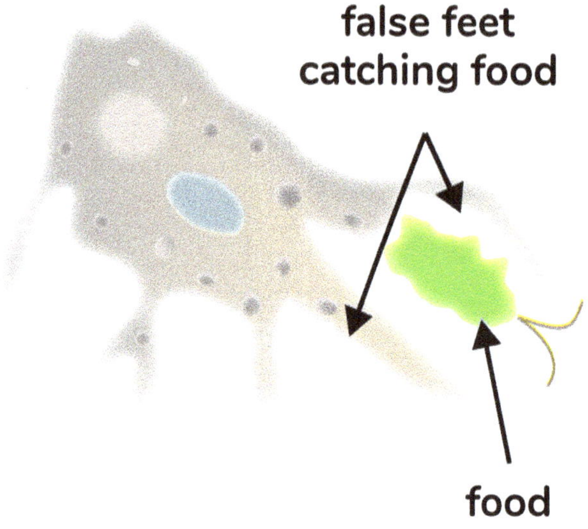

An amoeba uses its **false feet**, or **pseudopods**, to surround the food it wants to eat. Once the food is surrounded, the amoeba brings its feet together and makes something like a stomach out of the false feet that surround the food.

The stomach then **absorbs** the food into the body of the amoeba. That is how the amoeba eats with its feet!

8.4 Other Protists Eat

There are other protists that eat in other ways. For example, a protist called a ***Coleps*** rotates its whole body to swim through the water.

It also rotates as it eats. As it rotates, it uses its sharp teeth to bore through the food like a tiny drill. Then it eats the food it removes from the hole it has made with its teeth.

Coleps

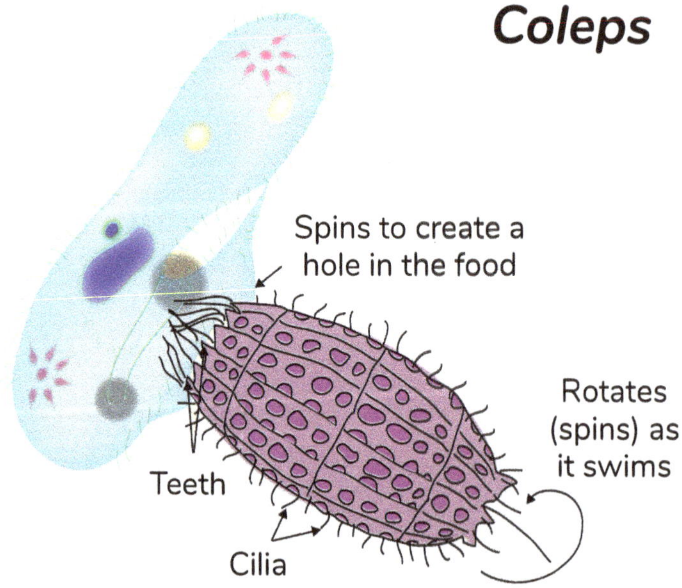

8.5 Summary

- A **euglena** uses **chloroplasts** to make its food.
- A **paramecium** uses **cilia** to swirl the water and sweep food into its mouth.
- An **ameoba** uses its **pseudopods**, or **false feet**, to capture food and eat.
- Other protists use other ways to eat.

8.6 Vocabulary

Basic Words

Coleps (KOH-lehps)

eyespot (IY-spaht)

Challenge Words

chlorophyll (KLAWR-uh-fihl)

chloroplast (KLAWR-uh-plaast)

paramecia (pehr-uh-MEE-shee-uh)

CHAPTER 9

Fungi: Molds, Mushrooms, YEASTS

9.1 Introduction

Have you ever eaten a **mushroom**? Some people love mushrooms, and other people don't like them at all.

Because mushrooms remind us of plants, we often think of mushrooms as being plants. Mushrooms grow in the ground, like plants, and can be found in the forest, like plants, but a mushroom is not a plant. A mushroom is a **fungus**. Mushrooms are different from plants because plants make their own food while mushrooms get their food by eating dead plants and animals.

There are three kinds of **fungi** — **molds**, **mushrooms**, and **yeast**.

Fungi (plural of fungus) have a whole kingdom, just like plants and animals do. All fungi are similar in some ways, and they all break down materials to get their food.

9.2 Molds

Molds are small fungi and are found everywhere. They can be in the soil, on wood, on dead plants, and on our food. Molds come in many different colors. They can be brown, white, orange, green, or black. Some molds are safe to eat, such as the mold in blue cheese, but many molds will make us sick. People can be allergic to molds that are around them, and some molds on food can cause illnesses.

Molds need water and food to grow, and they like to grow in warm places. Molds reproduce by making **spores**. Spores are too tiny to see and are usually made of just one cell. Given the right conditions, each spore can grow into a new mold. When we clean up the kitchen, we are cleaning up mold spores that we can't see. Cleaning up water and food stops molds from growing by removing their source of nutrients. We throw out bad food if it has mold on it because eating it could make us sick.

9.3 Mushrooms

Mushrooms grow in a variety of different shapes, sizes, and colors. Many mushrooms have a stem and a cap and look like tiny umbrellas.

Other mushrooms are shaped like cups and may or may not have a stem.

Mushrooms break down food and grow from their **hyphae**. Hyphae are very thin threads that spread out in the soil and are used by the mushroom to gather nutrients from the soil. In this way hyphae are similar to the roots of a plant. Hyphae are used by a mushroom to break down dead or decaying leaves, wood, or animals to use for food. By breaking down dead things, mushrooms help clean up and remove the dead things.

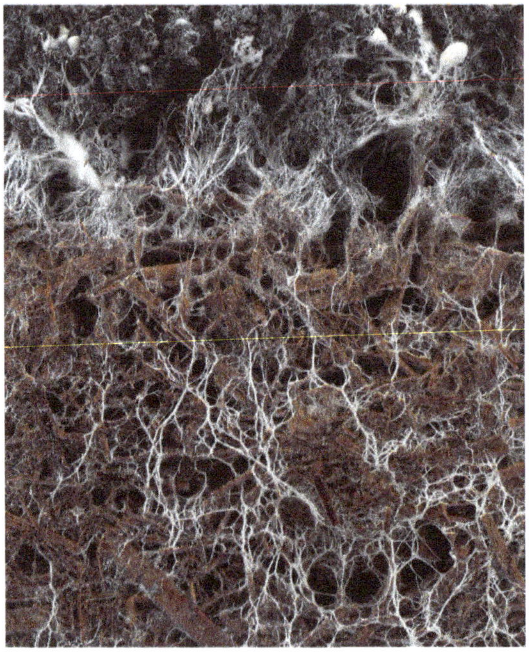

Like molds, mushrooms reproduce by making spores, and most mushroom spores are also too tiny to see.

Mushroom hunters are careful to gather only the kinds of mushrooms they know about because some mushrooms are very poisonous. You should not eat a mushroom that you have found outside of a grocery store unless a mushroom expert can tell you it is safe to eat.

9.4 Yeasts

Yeasts are fungi that we use in making bread, desserts, beer, and wine. Yeasts are some of the smallest fungi and are made of only one cell. A yeast reproduces by dividing its cell in two. This is called **budding**. Each of the two cells then grows until it is full size and can divide again.

Yeasts like to eat sugar. Like other fungi, yeasts like to be warm and moist and have lots of food. When you put yeast and sugar into warm water, the yeast will eat the sugar. As the yeast eats sugar, it creates **carbon dioxide** gas as a waste product. Yeast does this even when it has bread dough all around it! The carbon dioxide gas made by the yeast makes the bubbles you see in bread. The yeast dies when bread is baked. Oven temperatures are too hot for yeast.

9.5 Summary

- Some **fungi** help us, but other **fungi** ruin food and can make us sick.
- **Mushrooms** grow from **hyphae** which are root-like structures in the soil.
- **Molds** and **mushrooms** reproduce by making **spores**.
- **Yeasts** reproduce by **budding** — dividing one cell in half to make two cells.

9.6 Vocabulary

Basic Words

mold (MOHLD)

mushroom (MUHSH-room)

spore (SPAWR)

yeast (YEEST)

Challenge Words

fungus (FUHN guhs)

fungi (FUHN-jiy) or (FUHN-giy)

hyphae (HIY-fee)

CHAPTER 10

Physicist's TOOLBOX

10.1 Introduction

How do you study what happens when you throw a ball in the air? How do you find out why a marble rolls? How do you measure heat and what is happening when you rub your hands together and they get warm?

Physicists study things like motion, heat, energy, and power. Since these things are sometimes hard to see and understand, physicists use tools that help them measure, map, and record how the world around them works.

10.2 Brief History

Physics is considered by some scholars to be the oldest and most basic science. Ancient philosophers thought a great deal about how the world works.

One of these early philosophers was **Thales of Miletus** (625-545 B.C.E.). Thales was born in Greece, and he traveled to Egypt where he learned about geometry and astronomy.

Thales
625-545 B.C.E.

Ibn al Haytham
965-1040 C.E.

Greece was not the only place where philosophers were trying to understand how the world works. Ancient scholars in the Middle East, India, and China were all thinking about how the world works.

The Muslim philosopher **Ibn al-Haytham** studied light and how light rays travel.

Omar Khayyam was a Persian scientist who accurately calculated the length of a year.

Omar Khayyam
1048-1131 C.E.

In China, **Shen Kuo** was the first to describe the magnetic needle compass.

Shen Kuo
1031-1095 C.E.

10.3 Basic Physics Tools

Physicists study the world by measuring, mapping, and monitoring how things work. Physicists want to know things like how far away something is or how long it takes something to go from point A to point B. Physicists need to know **details**, and to find those details physicists measure distances, times, weights, and temperatures. Some of the **basic tools** physicists use are **timers**, **rulers**, **balances**, and **thermometers**.

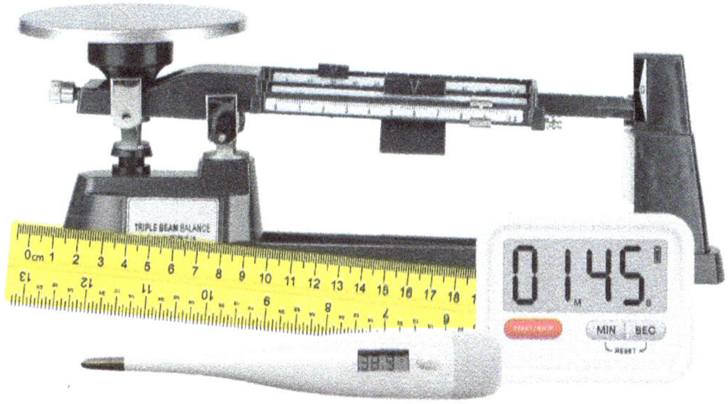

10.4 Advanced Physics

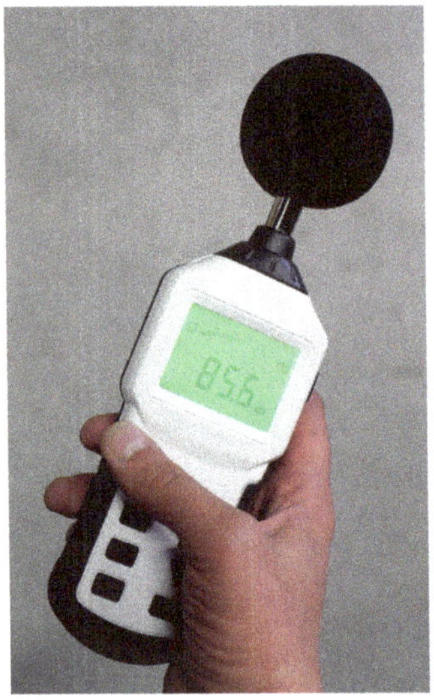

Physics covers a broad range of specialized subjects. Because there are so many subjects, there are many different kinds of tools.

Physicists who study sound might use a **sound meter** to measure high and low sounds.

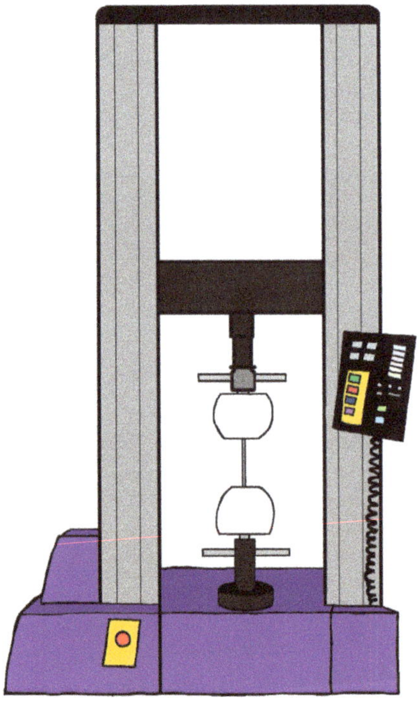

Physicists who study the properties of objects and materials might use different **dynamic testing machines** to find out more about what these things are like.

Some physicists study how objects go from hot to cold and might use a **thermal imager** to observe small changes in temperature.

Physicists who study electricity might use an **oscilloscope** to measure electrical waves.

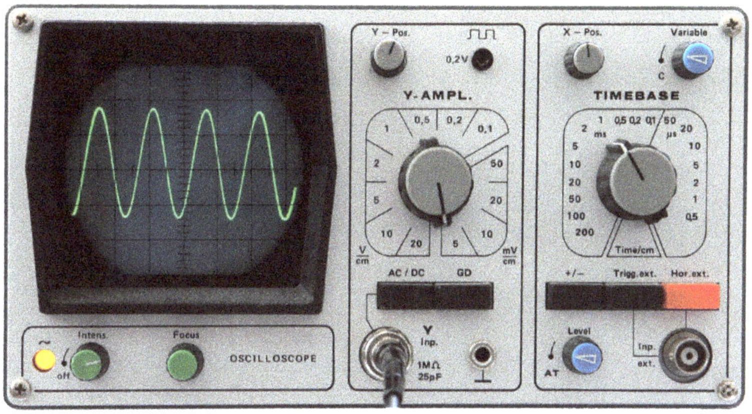

10.5 Computers and Robotics

Most advanced tools have computers inside or connect to a computer. In physics, computers are a very valuable tool. Computers can be programmed to collect large amounts of data. Computers can also be programmed to do complicated calculations or keep track of small distances.

Some high-level physics experiments use robots to move delicate instruments or record important information. Both computers and robotics have helped physicists learn more about how the world works.

10.6 Summary

- **Physicists** use a variety of **basic tools** and **advanced tools (instruments)** to do experiments.
- Many early philosophers from Greece, the Middle East, India, and China thought about how things work.
- Physicists use **basic tools** like **rulers**, **timers**, **balances**, and **thermometers**.
- Physicists use **advanced tools (instruments)** to measure **movement**, **temperatures**, **sound**, and **electricity**.
- **Computers** and **robots** help physicists make different kinds of measurements, collect large amounts of data, and do complicated calculations, among many other tasks.

10.7 Vocabulary

Basic Words

physics (FIH-zihks)

physicist (FIH-zuh-sihst)

Challenge Words

dynamic (diy-NAA-mihk)

Ibn al-Haytham (IHBN aal HAY-thuhm)

Omar Khayyam (OH-mahr kiy-AHM)

oscilloscope (ah-SIH-luh-skohp)

robotics (roh-BAH-tihks)

Shen Kuo (SHEHN GWOH)

Thales of Miletus (THAY-leez uhv miy-LEE-tuhs)

thermal imager (THUHR-muhl IH-mih-juhr)

thermometer (thuhr-MAHM-uh-tuhr)

CHAPTER 11

When THINGS MOVE

11.1 Moving Objects

A **force** is any action that changes the location or shape of an object or how fast or slowly an object is moving. You might think that forces also keep objects moving. This is what **Aristotle** thought. Aristotle was a philosopher in ancient Greece who studied how objects move. Aristotle thought that objects move because forces push on them from behind. However, 2,000 years later **Galileo**, an Italian physicist and astronomer, discovered that forces don't keep objects moving. Forces actually stop objects from moving. Forces can also change the direction of a moving object. But forces don't keep objects moving.

11.2 Keeping Objects in Motion

Galileo discovered that once an object is moving, it will keep moving unless it is stopped by a force. Forces can start an object moving or stop it, but forces do not keep an object moving. So, what keeps an object moving? An object keeps moving because of **inertia**. But what is inertia?

Have you ever been playing with your friends on the back porch when your mom called you to dinner? Maybe you didn't want to go to dinner. Maybe you wanted to keep playing. Maybe you even refused to stop playing and didn't go to dinner. That is inertia.

Inertia is when an object resists a change in motion.

When you kept playing with your friend and did not want to change what you were doing to go to dinner, you had inertia. You didn't want to change your motion (playing) to a new motion (going to dinner). When you resisted going to dinner, you were like an object that doesn't want to stop moving. That is inertia.

Inertia also keeps objects still.

When an object is moving, it wants to stay moving, and when an object is still, it wants to stay still. When an object does not want to change from moving to still or from still to moving, we say it has inertia.

11.3 Marbles and Bowling Balls

Everything has inertia. Bananas, oranges, bowling balls and marbles all have inertia. You have inertia. If you are standing in the middle of your room refusing to put your pajamas on, that is inertia. If you are running down a hill and you can't stop, that is inertia. So whether you are moving or not moving, you have inertia. Whether you are tall or short, young or old, you have inertia. Everything that has **mass** has inertia. But what is mass?

In physics, **inertia** comes from a property called **mass**. Everything has inertia because everything has mass.

Simply put, mass is how heavy something is. A heavy bowling ball has more inertia than a light marble. Because a bowling ball weighs more (has more mass) than a marble, it has more inertia.

Think about how hard it is to roll a bowling ball. Now think about how easy it is to roll a marble. It is harder to get a bowling ball to move than it is to get a marble to move. Why? Because a bowling ball has more inertia. It is also harder to get a bowling ball to stop moving than it is to stop a marble. Why? Again, because a bowling ball has more inertia than a marble.

11.4 Friction

We said that objects don't stop moving unless a force makes them stop. Because everything has inertia, an object in motion will stay in motion.

But wait! If you roll a soccer ball on the soccer field, it will eventually stop even if another soccer player doesn't make it stop. Why? Why does a soccer ball stop moving even if no one is there to make it stop?

It stops moving because of **friction**. Remember, all objects will keep moving (because all objects have inertia) unless a force makes them stop moving. Everywhere on Earth, friction makes things stop moving.

▶ **Friction** is a force that stops objects from moving.

Friction can come from almost anything. The grass in the soccer field makes friction that stops the rolling soccer ball. If a ball is thrown into a swimming pool, the ball will eventually stop moving because of the friction created by the water. Even air creates friction. If a ball is thrown into the sky, it will eventually slow down and stop moving because of friction caused by the air.

11.5 Summary

- A **force** is any action that changes the location or shape of an object or how fast or slowly an object is moving.
- **Forces** make objects start to move, and **forces** also stop objects from moving. But **forces** do not keep objects moving.
- **Inertia** keeps objects still and keeps objects moving.
- Simply put, **mass** is how heavy something is.
- **Friction** is a **force** that will stop objects that are moving.

11.6 Vocabulary

Basic Words

force (FAWRS)

friction (FRIHK-shuhn)

Challenge Words

Aristotle (EHR-uh-stah-tuhl)

Galileo (gaa-luh-LAY-oh)

inertia (ih-NUHR-shuh)

CHAPTER 12

Linear MOTION

12.1 Introduction

In this chapter we are going to look at a specific kind of motion called **linear motion**. Linear motion is simply motion that occurs along a line. If you push a hockey puck straight in front of you, it has linear motion. If you throw a baseball straight into the catcher's mitt, it has linear motion.

12.2 How Far?

Did you know that everything is moving? Even objects you think are standing still are moving. Your house is moving and even mountains are moving! You can tell houses and mountains are moving because you can see the Sun shine on them in the morning and then stop shining on them at night. Your house and the mountains around you are moving because the Earth is moving as it spins around and around on its axis. The Earth is also moving around the Sun.

Physicists say that **motion** is **relative**.

This means that we measure the motion of an object by comparing it to other objects. Physicists use the term **speed** to mean how fast an object moves. The speed of an object is *relative* to the objects surrounding it. For example, if a police officer pulls you over and hands you a speeding ticket, it is because you were moving too fast compared to him, the road, and the houses around him.

▶ The **speed** of an object is the measure of how far it goes in a given time.

If you start at your back porch and run to the back wall of your yard and it takes you three minutes, you ran that distance at a certain speed. But if your friend runs to the back wall of your yard and finishes the same distance in only one minute, then your friend ran faster than you did. A physicist would say your friend ran with more speed.

12.3 Average Speed

Physicists like to study details. If you ran to the back wall three times, a physicist would observe your speed the first time, the second time, and the third time you ran. The physicist would clock how fast you ran each time and would record three separate times. The speed at which you ran might have been affected by different things. For instance, if you got more tired each time you ran, your speed might have been slower the second time you ran and even slower the third time.

Imagine that you want to know on average how fast you can run to the back wall of your yard. To find your **average speed** you would add all three of your times together and then divide by three. This would tell you on average how fast you ran.

Knowing your average speed is useful information. You might find out that on average you can run to the back wall in two minutes, but on days when you feel really good you can run this distance on average in less than two minutes. By recording your average speed each day and making notes about how you were feeling and other things such as what you had for breakfast, you can compare your good days and your bad days and learn about what helps you run faster.

12.4 Acceleration

Imagine that you are running to the back wall and suddenly your friend starts to run beside you. You might try to win the race, so you speed up! Then as you get near the end, you might start to get tired and have to slow down.

When you started to run faster, you changed your **acceleration**.

> In physics, **acceleration** is the change in speed of a moving object for a given time.

For example, if you roll a marble down a ramp, you can watch it accelerate as it gets closer to the ground. The marble will start slowly and then gradually speed up. Depending on how long your ramp is, the marble might be moving very fast when it hits the ground. Because the marble speeds up as it travels down the ramp, physicists say it accelerates — changes its speed as it moves.

12.5 Summary

- **Linear motion** is the motion of an object in a straight line.
- **Speed** is how fast an object travels.
- **Average speed** is calculated by measuring the **speed** of an object several times, adding the speeds together, and dividing the total by the number of times the speed was measured.
- An object **accelerates** when it changes its **speed** while it is moving.

12.6 Vocabulary

Basic Words

speed (SPEED)

Challenge Words

accelerate (ihk-SEH-luh-rayt)

acceleration (ihk-seh-luh-RAY-shuhn)

average (AA-vuh-rihj)

linear motion (LIH-nee-uhr MOH-shuhn)

relative (REH-luh-tihv)

CHAPTER 12

Linear MOTION

12.1 Introduction

In this chapter we are going to look at a specific kind of motion called **linear motion**. Linear motion is simply motion that occurs along a line. If you push a hockey puck straight in front of you, it has linear motion. If you throw a baseball straight into the catcher's mitt, it has linear motion.

12.2 How Far?

Did you know that everything is moving? Even objects you think are standing still are moving. Your house is moving and even mountains are moving! You can tell houses and mountains are moving because you can see the Sun shine on them in the morning and then stop shining on them at night. Your house and the mountains around you are moving because the Earth is moving as it spins around and around on its axis. The Earth is also moving around the Sun.

Physicists say that **motion** is **relative**.

This means that we measure the motion of an object by comparing it to other objects. Physicists use the term **speed** to mean how fast an object moves. The speed of an object is *relative* to the objects surrounding it. For example, if a police officer pulls you over and hands you a speeding ticket, it is because you were moving too fast compared to him, the road, and the houses around him.

The **speed** of an object is the time it takes the object to travel a certain distance.

If you start at your back porch and run to the back wall of your yard and it takes you three minutes, you ran that distance at a certain speed. But if your friend runs to the back wall of your yard and finishes the same distance in only one minute, then your friend ran faster than you did. A physicist would say your friend ran with more speed.

12.3 Average Speed

Physicists like to study details. Assuming you ran exactly the same distance three times, a physicist would observe your speed the first time, the second time, and the third time you ran. The physicist would clock how fast you ran each time and would record three separate times. The speed at which you ran might have been affected by different things. For instance, if you get tired, your speed might slow down because you are taking longer to run the same distance.

Imagine that you want to know on average how fast you can run to the back wall of your yard. To find your **average speed**, assuming the distance is exactly the same, you would add all three of your times together and then divide by three. This would tell you on average how fast you ran.

Knowing your average speed is useful information. You might find out that on average you can run to the back wall in two minutes, but on days when you feel really good you can run this distance on average in less than two minutes. By recording your average speed each day and making notes about how you were feeling and other things such as what you had for breakfast, you can compare your good days and your bad days and learn about what helps you run faster.

12.4 Acceleration

Imagine that you are running to the back wall and suddenly your friend starts to run beside you. You might try to win the race, so you speed up! Then as you get near the end, you might start to get tired and have to slow down.

When you started to run faster, you changed your **acceleration**.

- In physics, **acceleration** is the change in speed of a moving object for a given time.

For example, if you roll a marble down a ramp, you can watch it accelerate as it gets closer to the ground. The marble will start slowly and then gradually speed up. Depending on how long your ramp is, the marble might be moving very fast when it hits the ground. Because the marble speeds up as it travels down the ramp, physicists say it accelerates — changes its speed as it moves.

12.5 Summary

- **Linear motion** is the motion of an object in a straight line.
- **Speed** is how fast an object travels.
- Assuming the distance stays exactly the same, **average speed** is calculated by measuring the time it takes an object to travel the stated distance and repeating several times. Then all of the time measurements are added together, and the total time is divided by the number of measurements taken. The result is the average speed.
- An object **accelerates** when it changes its **speed** while it is moving.

12.6 Vocabulary

Basic Words

speed (SPEED)

Challenge Words

accelerate (ihk-SEH-luh-rayt)

acceleration (ihk-seh-luh-RAY-shuhn)

average (AA-vuh-rihj)

linear motion (LIH-nee-uhr MOH-shuhn)

relative (REH-luh-tihv)

CHAPTER 13

Nonlinear MOTION

13.1 Introduction

In the last chapter we saw how the speed of an object can be measured when it is moving in a straight line. We learned that movement in a straight line is called linear motion. In this chapter we are going to look at **nonlinear motion**, or what happens when an object is not moving in a straight line.

13.2 Throwing a Ball

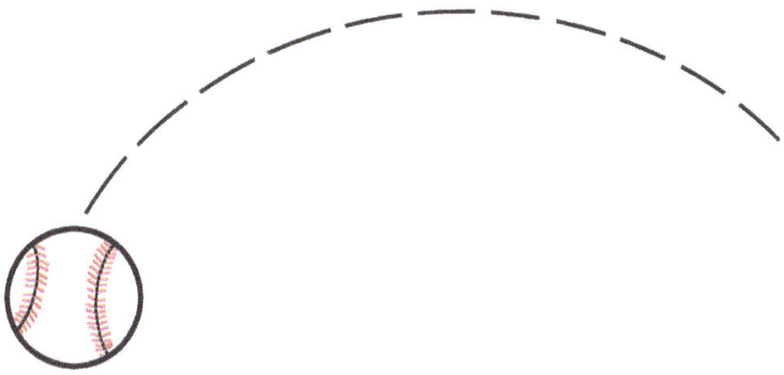

What happens if you go outside and throw a ball across your backyard? The ball moves forward, relative to your position, and travels to the other side of the yard. In the last chapter, we discovered that this type of motion is called **linear motion.**

But the ball not only moves forward, it also moves up and down. The up and down motion forms a curved path. Because this path is curved and not straight, it is called **nonlinear motion**. Nonlinear motion is motion that does not follow a straight line.

13.3 Riding a Bike

When you sit on a bike and push the pedals, you are using your legs to move you and your bike forward. If you are pedaling your bike and you look down at the frame of the bike, you will see that both you and the bike frame are moving forward in a straight line. This is an example of linear motion.

However, if you look below the bike frame and notice the wheels, you will see that the wheels are turning as you pedal.

In physics, we call this type of motion **nonlinear motion**.

The wheels are traveling in a circle, around the center hub. We could say the wheels are **rotating**. Also, because the wheels are traveling in an exact circle, they are moving with a special type of non-linear motion called **rotational motion**.

One way to observe the rotational motion of a bike wheel is to put a blue dot on the rear wheel and prop the bike up so the rear wheel is off the ground. As the pedals are pushed, the rear wheel moves, and the blue dot travels in a circle. This makes it easy to see the rotational motion of the wheel.

Bicycle wheels, car wheels, motorcycle wheels, and airplane propellers move with rotational motion. In the case of bicycles, cars, and motorcycles the rotational motion of the wheels is converted to linear motion as the vehicle moves forward. Wheels and propellers all are moving with rotational motion. They are all moving in a circle.

13.4 Easy and Hard Gears

Have you ever used a bike to go fast on a flat street? And have you ever used a bike to climb slowly up a steep hill? If you have a bike with more than one gear, you probably used a hard gear to go fast on a flat street and an easy gear to climb up a hill.

On a bicycle, **gears** are simple machines that use rotational motion to make it easier to climb hills and go fast on flat streets.

If you look carefully at the pedals of a single gear bicycle, you will see that they are attached to a metal wheel that has teeth. If you look at the rear wheel, you will see another metal wheel with teeth. Each of these metal wheels with teeth is called a **sprocket**. A chain goes around both of the two sprockets, joining them together.

When you use the pedals to turn the front sprocket, the chain transfers this motion to the sprocket on the back wheel, turning the wheel and moving the bicycle forward.

The combination of front sprocket, back sprocket, and the chain joining them is called a **gear**.

If you look at a bike that has more than one gear, you will see that it has several different size sprockets on the front and several different size sprockets on the back. By combining different sizes of front and back sprockets, you can make the bike easier or harder to pedal.

If the front sprocket is much larger than the back sprocket, the gear is harder to pedal but makes the back wheel move very fast. On a flat surface this gear is used for speed.

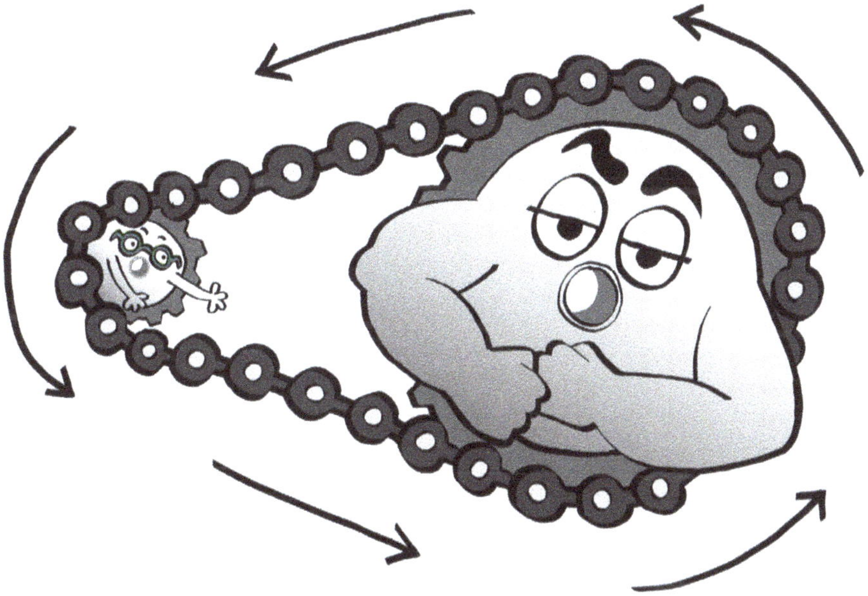

If the front sprocket is smaller than the back sprocket, the gear is easier to pedal but moves the back wheel very slowly. Because the gear is easier to pedal, this gear can be used to climb hills.

By understanding how gears work, you can pick gears to help you go uphill with less effort and downhill faster!

13.5 Summary

- **Nonlinear motion** is motion of an object that moves in a curve or in a circle.
- **Rotational motion** is a type of non-linear motion of an object that **rotates** or moves in a circle.
- Bicycles, cars, motorcycles, and airplane propellers use **rotational motion** of the wheel or propeller to move them forward.
- The **gears** on a bicycle use **rotational motion**. Different size **sprockets** are used to make a bike go faster or make it easier to climb hills.

13.6 Vocabulary

Basic Words

gear (GEER)

sprocket (SPRAH-kuht)

Challenge Words

nonlinear (nahn-LIH-nee-uhr)

rotational (roh-TAY-shuh-nuhl)

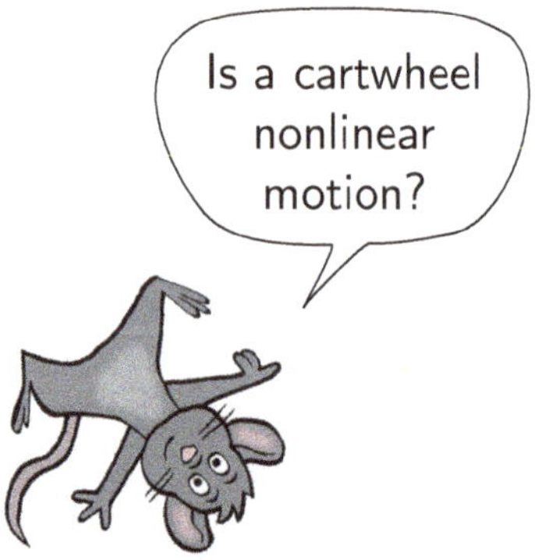

CHAPTER 14

Geologist's TOOLBOX

14.1 Introduction

Geologists are scientists who use special tools to study Earth. Some tools are basic, like hammers and lenses, and some tools are more advanced, like seismographs and submarines.

There are different tools for different types of geologists. Some geologists study what things are made of and may use test tubes and chemicals to test minerals and rocks. Some geologists study how things are put together and might use hammers and microscopes to examine layers of sediments and inside rocks.

Because there are many different types of geologists, there are many different types of tools in a geologist's toolbox.

14.2 Brief History

It's difficult to know exactly what the first geology tool was and when it was invented. It is clear from a variety of ancient texts that early geologists were curious about what was inside rocks and beneath layers of dirt, and they likely used simple tools such as hammers and shovels to examine rocks and minerals.

Theophrastus was a student of Aristotle and wrote a book called *On Stones* in which he describes the color, hardness, and smoothness of a variety of both common and rare stones. Many of these stones were found on the ground, in streams, and by mining.

Theophrastus
372–287 B.C.E.

It is also clear from writings of Theophrastus that he explored the nature of stones by performing various experiments. By using fire, he could observe which stones melted or changed color when burned. Theophrastus also used acids to examine which stones would dissolve and which would withstand these chemical tests.

14.3 Basic Tools: Hammers and Lenses

Three types of **rock hammers** used by geologists are: the **pointed tip rock hammer**, the **chisel edge rock hammer**, and the **crack hammer**.

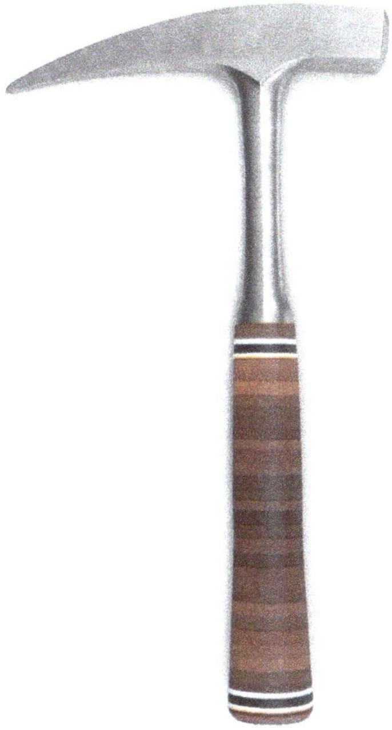

A **pointed tip rock hammer** has one end that comes to a point. The other end has a head that is flat and square. Pointed tip rock hammers are also called **rock picks**, and they are generally used by geologists when they are working with hard rocks. Geologists often use the pointy tip to dig fossil samples out of rocks and the square end to crack open a rock to see what's inside.

A **chisel edge rock hammer** has one end that is flat and broad like a chisel rather than being pointy. The other end is a square head. The chisel end is used to split layers of soft rock. The square head is used to crack open rocks.

A **crack hammer** has two blunt ends. It is usually heavier than either a pointed tip rock hammer or a chisel edge rock hammer. Its heaviness can make it easier for geologists to break rocks open.

Once a geologist breaks a rock open with a rock hammer, is there a way to get a better look at what's inside? To better see the features of the inside of a rock, geologists often use a **hand lens**. A hand lens is a small magnifier that folds into a holder, making it compact and easy to carry. A hand lens can be carried in a pocket or on a string around the neck.

A hand lens magnifies a rock, making it easier to see. With a hand lens, a geologist can look at the details of a rock. This can help determine what the rock is made of.

14.4 Advanced Geology Tools

Modern geologists use a variety of advanced tools to study Earth's rocks, minerals, soils, and Earth's activity. **Computers**, **satellites**, **radios**, and **other modern technologies** play a very important role in how geologists collect and analyze data.

A **global positioning system** or **GPS** is an electronic device that has greatly changed how geologists map and study Earth's surface. A GPS device receives signals from a network of satellites orbiting the Earth. The GPS unit then uses this data to calculate its location on the Earth. Many GPS devices are small enough to be easily tucked into a backpack or pocket. Geologists use GPS devices to help them do things such as navigate, create geographical maps, locate specific landmarks, and determine property lines.

Another advanced technology that has helped geologists map Earth's surface is the **geographic information system** or **GIS**. A GIS is a system that uses computers to gather and sort different types of geographical data. The data can then be used to create maps that help geologist examine mountains, rivers, volcanoes, earthquakes, and even geological features beneath the ocean!

14.5 Summary

- Geologists use both **basic and advanced tools**.
- Many geologists use **basic tools** such as **rock hammers** and **hand lenses**.
- Geologists also use **advanced electronic tools** such as the **GPS (global positioning system)**.
- **Computers** play an important role in the study of geology. Geologists use **computer systems**, such as **GIS (geographic information system)**, to study Earth.

14.6 Vocabulary

Basic Words

rock hammer (RAHK HAA-muhr)

Challenge Words

computer (kuhm-PYOO-tuhr)

electronic (ih-lehk-TRAH-nihk)

geologist (jee-AH-luh-jihst)

geology (jee-AH-luh-jee)

geographic information system [GIS]
(jee-uh-GRAA-fihk ihn-fuhr-MAY-shuhn SIHS-tuhm)
[JEE IY ESS]

global positioning system [GPS]
(GLOH-buhl puh-ZIH-shuh-nihng SIHS-tuhm)
[JEE PEE ESS]

satellite (SAA-tuh-liyt)

technology (tehk-NAH-luh-jee)

Theophrastus (thee-uh-FRAAS-tuhs)

CHAPTER 15

Earth's SPHERES

15.1 Introduction

If you go outside and walk around in your backyard, you will notice a variety of features that make up the place on Earth where you live.

Geosphere

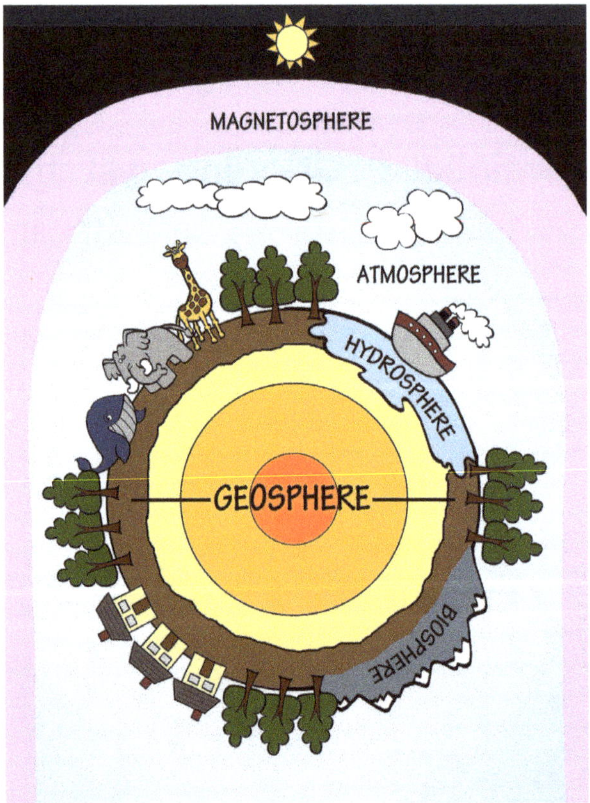

When you jump up and down, you can feel the hard rock surface below your feet. If you have a shovel, you might dig a little way below the surface and observe dirt, rocks, and minerals. This is the "rock part" of Earth, called the **crust**. The crust and all of the layers of Earth below the crust together are called the **geosphere**.

Atmosphere

When you inhale and exhale, you feel your lungs fill with oxygen and then release carbon dioxide. This is the "air part" of Earth called the **atmosphere**.

Biosphere

If you walk on the grass or kneel down to examine tiny ants taking off with crumbs from your peanut butter sandwich, you are looking at the "biology part" of Earth called the **biosphere**.

Hydrosphere

If it starts to rain on you while you are looking at the ants scurrying on the ground and carrying your sandwich crumbs, you are experiencing the "water part" of Earth called the **hydrosphere**.

Magnetosphere

Finally, if you use a compass to find your way home, you are interacting with the "magnetic part" of Earth called the **magnetic field,** which is within the **magnetosphere**. The magnetosphere is the area surrounding Earth in space where Earth's magnetic field interacts with gases given off by the Sun.

The Magnetosphere

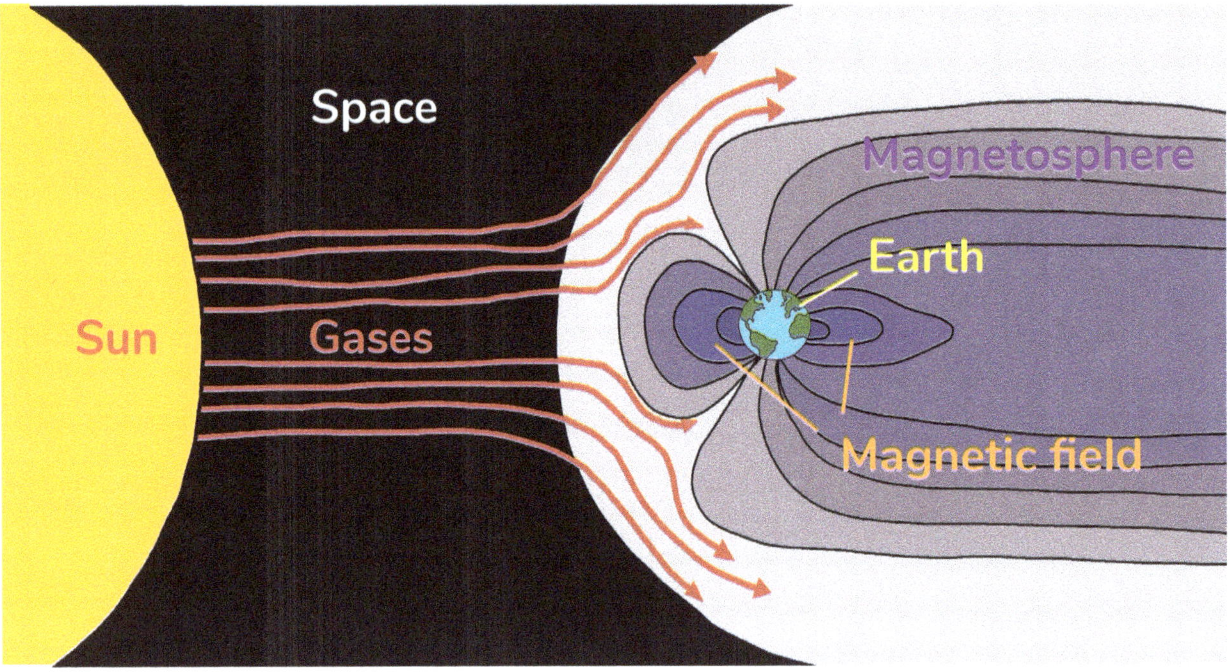

The geosphere, atmosphere, biosphere, hydrosphere, and magnetosphere can be studied both separately and together. As a whole, they make up this planet we know as Earth.

15.2 Why Spheres?

Notice that all of these names end in the term **sphere**. The word sphere refers to a ball-shaped object. Earth is a ball-shaped or **spherical** object in space.

When geologists talk about any of the "spheres" of Earth, they are talking about a whole part, or layer, that encircles the Earth.

The geosphere refers to all of the rock and mineral layers that make up Earth's surface and interior. The atmosphere is the whole layer of air that covers Earth. The biosphere contains all the living things that inhabit Earth. The hydrosphere has all the water that is found on Earth, and the magnetosphere includes the entire magnetic field of Earth.

When geologists study a particular part of a layer of the Earth, they use words that describe specific features. For example, mountains, rocks, and the ocean floor are different parts of the geosphere. Oceans, lakes, and glaciers are different parts of the hydrosphere. A **biome** is a particular part of the biosphere in which the climate, soils, and plant life are similar throughout.

15.3 Why Study Earth's Spheres Separately?

Sometimes it helps to understand how something works by taking it apart and studying each piece separately. If you want to know more about how your bicycle works, you could take off the wheels and the tires, take off the chain, and remove the sprockets to study each part by itself.

You might discover that your bicycle frame is built out of sturdy metal and that it is hollow inside to make it lighter. You might discover that your tire tubes are made out of soft rubber and the tires are thicker with small bumps or treads that help the tires grip the road. You might also discover that the chain is flexible and can bend and that a sprocket is hard and has teeth to support and carry the chain.

By studying each piece of your bike separately, you can better understand how your bike works when it's all put together.

The same is true about studying Earth. It's useful for geologists to look carefully at each sphere and learn everything they can about water or air or the rocks that make up Earth. Also, because there is so much to learn, no one person can study everything. This is one reason geologists specialize in different areas of geology. A **geochemist** may focus only on studying the water in the oceans, and a **structural geologist** may only study the mountains beneath the oceans. When they work together and swap information, they both understand the oceans better.

15.4 Putting Them All Together

Just like your whole bike functions as a combination of different parts working together, Earth operates as a combination of different parts that work together. The atmosphere, biosphere, geosphere, hydrosphere, and magnetosphere all work together to create the planet we call Earth.

For example, when the atmosphere creates a big storm over the ocean, the hydrosphere is affected, and the biosphere and geosphere may also be affected. When animals in the biosphere, like beavers and people, build dams, it can affect the geosphere, hydrosphere, and other parts of the biosphere. When a volcano erupts, it can affect the geosphere, atmosphere, and biosphere.

By studying the individual spheres separately and then pooling all that information to study Earth as a whole, geologists can gain a more accurate picture of this world we call Earth.

15.5 Summary

- Earth can be studied as a combination of different **spheres**.
- The five major spheres that make up Earth are the **geosphere**, **atmosphere**, **biosphere**, **hydrosphere**, and **magnetosphere**.
- Studying each **sphere** separately helps scientists learn more about how they work.
- Combining the information about each **sphere** into the study of Earth as a whole helps scientists gain a more accurate picture of Earth.

15.6 Vocabulary

Basic Words

sphere (SFIHR)

Challenge Words

atmosphere (AAT-muh-sfihr)

biosphere (BIY-uh-sfihr)

geochemist (jee-oh-KEH-mihst)

geosphere (jee-oh-sfihr)

hydrosphere (HIY-droh-sfihr)

magnetic (maag-NEH-tihk)

magnetosphere (maag-NEE-tuh-sfihr)

spherical (SFIHR-ih-kuhl)

CHAPTER 16

The GEOSPHERE

16.1 Introduction

The term **geosphere** refers to the part of Earth that is made of rocks, minerals, and soils. The geosphere extends from the surface of Earth all the way to the very center. Geological activity occurs in the geosphere and includes earthquakes and volcanoes.

Scientists think the geosphere is made of several layers that are different from one another. One way to describe the geosphere is to divide it into three main layers: the **crust**, the **mantle**, and the **core**.

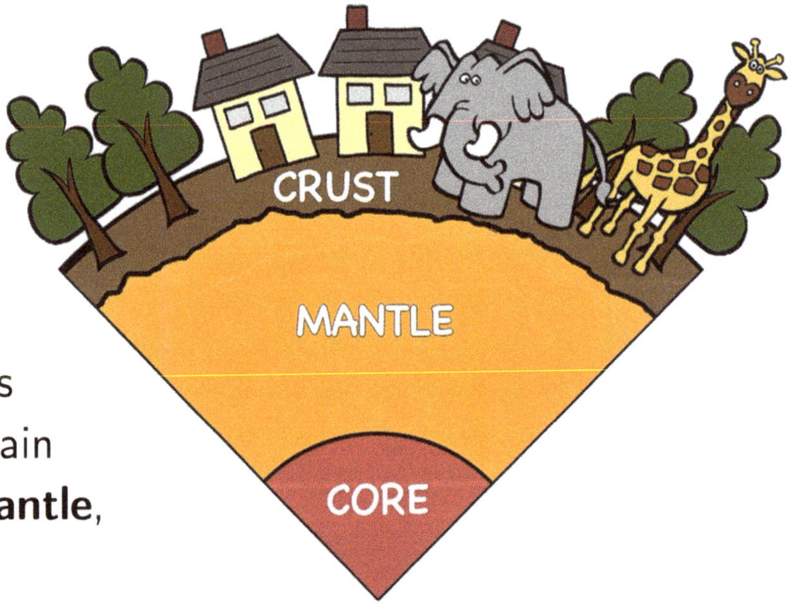

To create a more detailed picture of Earth's layers, geologists often subdivide the main layers of Earth into additional layers. This is helpful when more specific descriptions of the geosphere are needed. Scientists group and name the various layers in different ways according to what will be most helpful to them in studying the geosphere. There is no one right way to list Earth's layers. It just depends on what kinds of questions scientists want to answer.

16.2 More About Layers

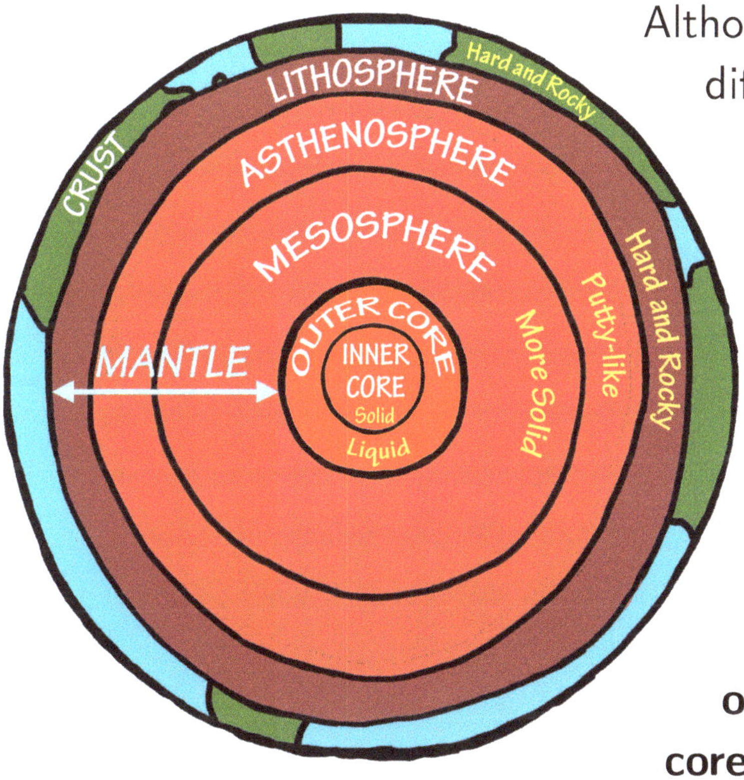

Although scientists can use different names for the layers of the geosphere, in this book we will use the following names for the layers: the **crust**, the **mantle** (divided into the **lithosphere**, **asthenosphere**, and **mesosphere**), and the **core** (divided into the **outer core** and the **inner core**).

The **crust** is the outermost layer of Earth and is the part of the geosphere that we live on. The crust is hard and is made of rocks, soil, and minerals. The crust makes up mountains, the ocean floor, mesas, river beds, and other features.

- Below the crust is the **mantle**, which is further subdivided into three layers called the **lithosphere**, **asthenosphere**, and **mesosphere**.

The **lithosphere** lies just below the crust and, like the crust, is a hard, rocky layer. Geologists believe the lithosphere is broken up into huge pieces called **plates**. It is thought that the lithosphere sits on top of a soft, putty-like layer called the **asthenosphere** which is made of molten (melted) rock called **magma**. Heat in the asthenosphere causes the magma in this layer to move. As the magma moves, it carries the plates of the lithosphere along with it, causing earthquakes and volcanoes as the plates shift and push against each other.

Below the asthenosphere is the **mesosphere**. The mesosphere makes up the largest part of the mantle. Scientists believe that the mesosphere is more solid than the asthenosphere.

- Below the mesosphere and at the very center of Earth is the **core**.

The core is divided into the **outer core** and the **inner core**. The outer core is thought to be made of liquid rock and metal, and the inner core is thought to be solid.

Since it isn't possible to drill to the center of the Earth, geologists don't know exactly what the outer core and the inner core are like, but they can make **educated guesses** based on the data they collect from experiments.

It's good to keep in mind that scientists sometimes use different groupings and names for the layers of Earth. For example, sometimes the lithosphere and asthenosphere are grouped together and called the **upper mantle** with the mesosphere being called the **lower mantle**. And at other times, geologists may group the crust and lithosphere together into a layer they simply call the lithosphere.

16.3 Evidence and Guessing

There is much about the geosphere that geologists simply don't know. Because we have not been able to dig below the Earth's crust, we can't get samples of the Earth's interior. This means that some of the conclusions scientists make about how the geosphere works are more like **educated guesses**.

An educated guess is a guess based on scientific information. When there is enough information to suggest that an educated guess is correct, the guess can become a scientific theory. Sometimes an educated guess is discarded when new evidence suggests that the idea is incorrect. Either way, we gain more understanding of how the geosphere works by looking at evidence, making educated guesses, developing theories, and then discarding theories and guesses when new information challenges old ideas.

For example, as we saw earlier, geologists have found evidence that supports the idea that the lithosphere is divided into plates and earthquakes are caused by the movement of these plates on the soft asthenosphere. Scientists also think that the magma in the asthenosphere can be forced through cracks or thin places in the lithosphere, causing volcanoes to erupt.

There is no way to actually sample the lithosphere or the asthenosphere, so geologists don't know for certain that the lithosphere is hard and the asthenosphere is soft and putty-like. However, they can use advanced tools to gather information, and based on that information they can make an educated guess about what the lithosphere and the asthenosphere are like.

16.4 The Geosphere and Other Spheres

The crust is one part of the geosphere that is affected directly by the other spheres, like the atmosphere, the biosphere, and the hydrosphere. The crust is shaped both by the lower layers of Earth where earthquakes and volcanoes begin and by wind, rain, storms, and animals.

For example, a volcano might erupt and create a tall mountain. Over time the mountain may erode and become smaller and smaller because of wind and rain. An earthquake may occur, creating a new passageway for water, and animals might build their homes in the new waterway, creating small ponds or lakes.

The magnetic field that surrounds Earth is believed to be created by the motion of liquid metals in the outer core. If Earth didn't have a core made mostly of metals, we would not have a magnetic field to form the magnetosphere.

The lower layers of the geosphere are not directly affected by the other spheres of the Earth. It doesn't rain on the asthenosphere, and animals can't dig homes in the mesosphere.

16.5 Summary

- The **geosphere** includes all the rock parts of Earth and extends from the surface of Earth all the way to the very center.
- The **geosphere** can be divided into three main layers: **crust**, **mantle**, and **core**. These main layers can be further subdivided into more layers.
- Geologists can only take samples of the outermost layer of the **geosphere** (the **crust**) and have to make **educated guesses** about the inner layers.
- The **crust** is the layer of the geosphere most affected by other spheres.

16.6 Vocabulary

Basic Words

core (KAWR)

crust (KRUHST)

magma (MAAG-muh)

mantle (MAAN-tuhl)

Challenge Words

asthenosphere (aas-THEH-nuh-sfihr)

educated guess (EH-juh-kay-tuhd)

lithosphere (LIH-thuh-sfihr)

mesosphere (MEH-zuh-sfir)

CHAPTER 17

The Air We BREATHE

17.1 Introduction

Notice what happens when you breathe. Take a deep breath in. What is going into your lungs? Exhale. What comes out?

Air is the name for what we breathe in and what we breathe out. We live on a planet that has the kind of air needed for life. Without the air on Earth, there wouldn't be animals on farms, ponds full of fish, or forests full of trees.

17.2 The Atmosphere

The air we breathe exists in the part of the Earth that scientists call the **atmosphere**. The atmosphere sits just above the Earth's crust and extends for several miles above the surface. Earth is the only planet in our **solar system** that has an atmosphere suitable for life as we know it.

Most of the time we don't think too much about the air that surrounds us. We breathe it in and we exhale it out as we go about our day. Some people, though, live in areas where the air is not clean. In some areas of the world, the air contains so much **pollution**, or small particles, that it is difficult to breathe. Knowing about the air, what it is, and how to keep it clean is important to all living things on Earth.

17.3 What Is Air?

You might think most of the air we breathe is **oxygen**. But it isn't. The air we breathe is a mixture of different **gas molecules** and **water vapor**. Air has **nitrogen gas**, **oxygen gas**, **carbon dioxide gas**, a little bit of **argon gas**, and some **water vapor**.

Nitrogen Gas Molecule

Nitrogen gas in the air exists as a molecule with two nitrogen atoms hooked together. When you breathe, nitrogen gas goes inside your lungs and comes out again when you exhale. Your body doesn't use nitrogen gas. It only uses oxygen gas.

You have probably heard that you need **oxygen** to stay alive. In fact, if you hold your breath too long, you will pass out and your body will automatically start breathing again to keep you alive. Oxygen is more important than food or water, and you can't stay alive very long without it.

Oxygen Gas Molecule

Oxygen gas is made of two oxygen atoms hooked together. When you breathe in, your lungs expand and oxygen gas goes inside.

Lungs have special cells that absorb oxygen from the air and deliver it to your blood. Your blood has special molecules that then carry the oxygen to the rest of your body. Your body uses the oxygen to process food and give you energy.

After our bodies use the oxygen we have inhaled, they make **carbon dioxide gas**. Carbon dioxide gas is made of one carbon atom and two oxygen atoms hooked together. When you exhale, you breathe out the carbon dioxide gas.

Carbon Dioxide Gas Molecule

This works out great because plants use carbon dioxide to make the food that allows them to stay alive and grow. Plants take in carbon dioxide and then put oxygen back into the air. We help plants and plants help us!

17.4 Why Doesn't Air Float Away?

You might have wondered if there is oxygen, nitrogen, or carbon dioxide in space. In fact, when astronauts travel outside our atmosphere and into space, they have to wear a special suit so they can have oxygen to breathe.

But what keeps the air we breathe close to the Earth, and why doesn't it just all float away into space? In fact, what keeps you on the Earth, and why don't you float away?

You don't float away for the same reason the air in our atmosphere doesn't float away. **Gravity** keeps you on the ground and keeps the air we breathe close to the Earth's surface. Gravity is a force that pulls everything near the Earth toward its center. Go ahead and try to jump off the Earth. As soon as you jump up, you will feel Earth's gravity pulling you back down again.

17.5 Summary

- The **air** we breathe is in the Earth's **atmosphere**.
- The air we breathe is made up of **nitrogen gas**, **oxygen gas**, **carbon dioxide gas**, **argon gas**, and **water vapor**.
- Our bodies inhale air, use the **oxygen**, and then exhale **carbon dioxide gas**.
- **Gravity** keeps the air from floating away into space.

17.6 Vocabulary

Basic Words

argon (AHR-gahn)

solar (SOH-luhr)

vapor (VAY-puhr)

Challenge Words

carbon dioxide (KAHR-buhn)

gravity (GRAA-vuh-tee)

nitrogen (NIY-truh-juhn)

pollution (puh-LOO-shuhn)

CHAPTER 18

Astronomer's TOOLBOX

18.1 Introduction

How can you study objects in space when they are so far away? How can you make discoveries about what other solar systems are like when you can't get on a spaceship and go to them? How can you find out what other planets are like when you can't land on them?

Astronomers depend on a variety of tools to help them visualize the cosmos. New discoveries in astronomy have been made possible by the invention of **computers**, **satellites**, and **spaceships** that can travel far from Earth. Although the study of astronomy is changing quickly as more new tools help astronomers explore space, the invention of a **basic tool**, the **telescope**, was the first step in the exploration of space.

18.2 Brief History of a Basic Tool

In the 1400s and 1500s when **Copernicus** was gazing at the night sky, looking up at the stars and wondering about the cosmos, he had to do all of his investigating without the help of a **telescope**.

- Copernicus was smart, and he revolutionized the way we think about the universe by using mathematics to show that the Earth moves around the Sun and that the Sun remains fixed in one location. But without being able to see the details of the Sun and of planets and their moons, Copernicus was limited in what he could explore.

Galileo was exploring the skies, just like Copernicus, but Galileo was born during the time when the telescope was being invented. In the early 1600's Galileo found out about a **spyglass** invented by a Dutch lens maker named **Hans Lippershey**. Galileo knew that this spyglass could help him study the cosmos by allowing him see things that were far away in more detail.

- Galileo used Lippershey's ideas to create his own spyglasses that would eventually become telescopes.

18.3 What Is a Telescope?

A **telescope** is a tool or **instrument** that helps astronomers see far into the distance. A simple telescope has two **lenses** connected by a long tube.

One lens is called the **eyepiece** and is located at one end of the tube. You look through this lens with your eye.

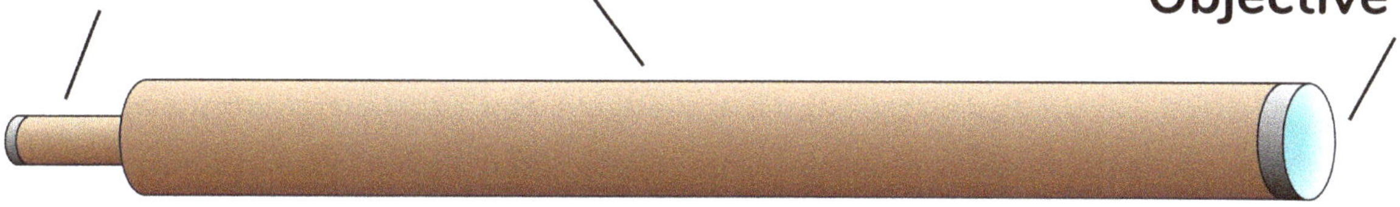

The other lens is called the **objective lens** and is located at the other end of the tube. The objective lens is used to collect light so an object may be viewed.

As light travels through the objective lens and down the tube, it is focused by the eyepiece. In this way the object being viewed becomes **magnified**, or made to appear larger. The larger the objective lens, the more light can be collected, and the longer the tube, the larger the object will appear.

18.4 Early Telescopes

Early telescopes looked very much like the simple telescope described in the previous section. Early telescopes were often made of metal, such as brass or copper, and had two simple lenses, one at either end of a long tube.

Early Telescope

The body of the metal tube was sometimes attached to a stand. The astronomer could pivot the telescope on its stand, adjusting the direction in which the telescope was pointing.

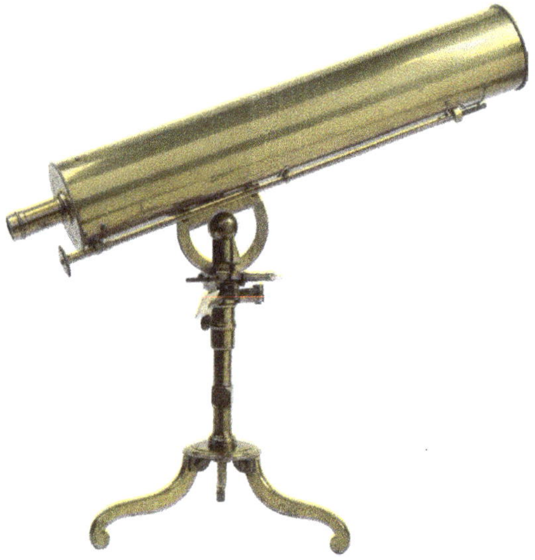

Gregorian Telescope

Part of the problem with these early telescopes was that in order to see a faraway object the tube had to be very long. With the invention of the **Gregorian telescope** and the **Newtonian telescope**, mirrors were added to the inside of the telescope tube to solve this problem.

In both the Gregorian and Newtonian telescopes, mirrors in the tube are used to focus the light that enters through the objective lens. When the light hits the mirrors, it is bounced from the mirrors into the eyepiece. This design allows astronomers to see distant objects with telescopes that have shorter tubes.

18.5 Advanced Telescopes

The telescope has come a long way from the spyglass of Galileo's day. Technological advances have resulted in significant improvements for the modern telescope. Today, astronomers can see not only the planets and their moons in our own **solar system** but also planets and moons in other solar systems!

A **galaxy** is a large group of solar systems, stars, and other objects in space. Using modern telescopes, astronomers can now see whole solar systems, not only in Earth's galaxy, but in other galaxies as well!

The **Hubble Space Telescope** is an **advanced telescope**. As the name suggests, the Hubble Space telescope is actually in space orbiting the Earth. The Hubble Space Telescope was launched into space in 1990 and has produced some of the most amazing images of space we have ever seen.

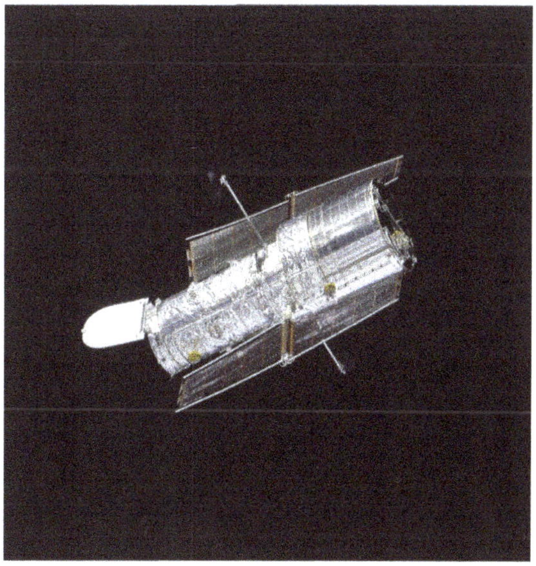

18.6 Summary

- Astronomers explore space by using **tools** such as **telescopes**, **computers**, **satellites**, and **spaceships**.
- **Galileo** modified the **spyglass** to create a **telescope**.
- Early **telescopes** had an **objective lens** and an **eyepiece** that were connected by a long tube.
- The **Hubble Space Telescope** is an **advanced telescope** that is in space and travels around the Earth.

18.7 Vocabulary

Basic Words

spaceship (SPAYSS SHIHP)

Challenge Words

Copernicus (koh-PUHR-nih-kuhs)

Gregorian (grih-GAW-ree-uhn)

Hubble Space Telescope
(HUH-buhl SPAYSS TEH-luh skohp)

Hans Lippershey (HAHNZ LIH-puhr-shee)

instrument (IHN-struh-muhnt)

Newtonian (noo-TOH-nee-uhn)

objective lens (uhb-JEHK-tihv LEHNZ)

satellite (SAA-tuh-liyt)

telescope (TEH-luh skohp)

CHAPTER 19

Observing the CONSTELLATIONS

19.1 Introduction

No matter where you are on Earth, if you look up to the sky on a clear, dark night, you can see stars. If you are far away from city lights, it looks like the sky is filled with thousands and thousands of stars. There are so many stars that it is difficult to know the name of each star in the sky.

Even without using a telescope, modern satellites, or space probes, we can learn something about the cosmos by observing stars that form **constellations**.

A constellation is a group of stars that together appear to form a shape or image in the sky. By grouping stars into constellations, a lot can be learned about the sky without knowing all the individual stars.

Some common constellations easily observed in the Northern Hemisphere include the **Big Dipper** and **Cassiopeia**. Favorite Southern Hemisphere constellations include the **Archer** and the **Whale**.

Since Earth is a large sphere, the stars someone can see at the North Pole are different from the stars someone can see at the South Pole. If you are at the equator, over the course of a year you can see all the constellations, but the stars at the North and South Poles will be harder to see.

The constellations change their positions in the sky throughout the night and during the different seasons. This repositioning of the stars is caused by the Earth changing its position in space as it spins on its axis and revolves around the Sun. As the Earth moves, we see the constellations from different angles.

By observing the positions of the constellations in the sky, you can watch the seasons come and go, tell what time of night it is, and determine which direction you are going when traveling on land, sea, or in the air. Knowing the constellations could even help you find your way home if you were taking a trip in outer space!

19.2 Northern Hemisphere Constellations

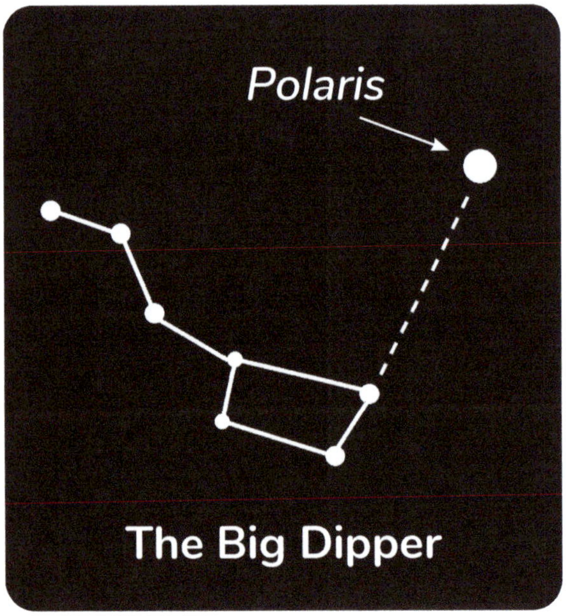

If you live in the **Northern Hemisphere** there are many constellations you can easily find. A favorite group of stars in the northern sky is called the **Big Dipper** because its shape looks like a dipper, or ladle. The Big Dipper has 7 stars, three of which form the "handle" and four that form the "bowl" of the dipper. The best time to see The Big Dipper is from February through June.

The Big Dipper can be used to locate the star **Polaris**, which is also called the **North Star**. Polaris can be found by imagining a line going between the two stars located at the end of the bowl of the Big Dipper and extending the line to a lone star that has no other stars nearby. This star is Polaris.

Polaris is above the northernmost point of the Earth. Because it is near enough to true north, it can be used as a navigational marker. If you were on a boat sailing from Maine to England on a clear night, you could use Polaris to keep your boat on track.

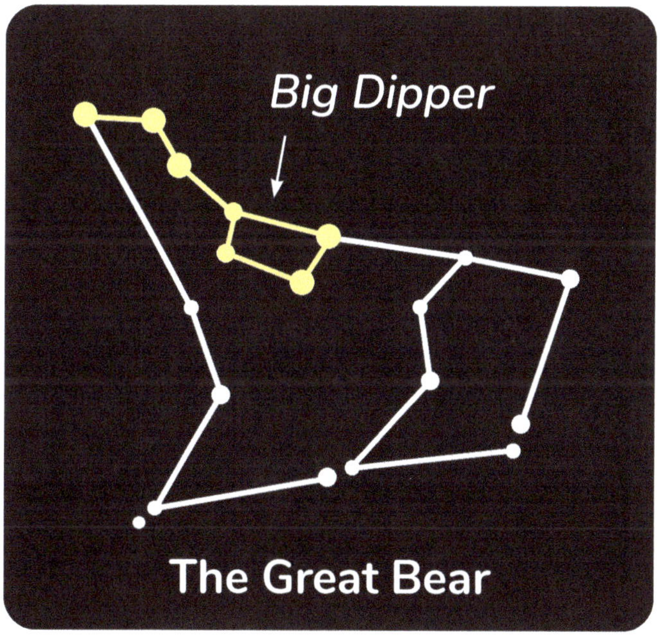

Another interesting constellation to see in the Northern Hemisphere is the **Great Bear**. The Great Bear includes the Big Dipper. To locate the Great Bear, first find the Big Dipper, and while keeping the Big Dipper in view, expand your gaze to include the three pairs of stars that form the Bear's paws. Once you have these stars and the Big Dipper in view, you can see the rest of the stars that make up the Great Bear. The best time to see the Great Bear is from February through June.

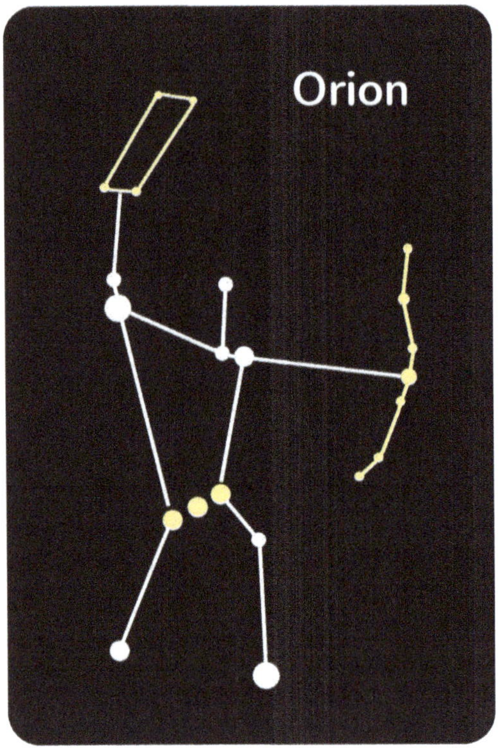

An easy to find constellation is **Orion the Hunter**. The stars that make up Orion are bright and beautiful which makes it easy to pick out this constellation. A good way to find Orion is to find his belt. Orion's belt is made of three bright stars close together in a straight row. From there you can pick out the shield and raised club. The best time to see Orion is December through March.

19.3 Southern Hemisphere Constellations

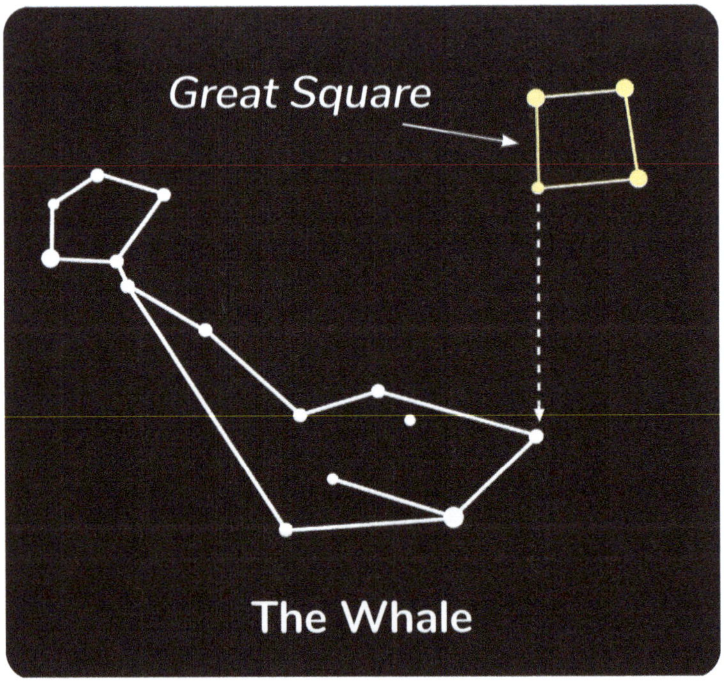

If you live in the Southern Hemisphere, there are lots of fun constellations you can find. One of the largest southern constellations is the **Whale**. The stars that make up the Whale are dim, but because there are fewer stars to observe in this section of the sky, the Whale can be easily seen on a dark night when there is no Moon or city lights.

To find the Whale, first locate the **Great Square**.

As the name suggests, the Great Square is a set of four bright stars that form a square. Once you locate The Great Square, the Whale is easy to spot. Just follow the line made by the two stars on the side of the Great Square until you see a cluster of 4 to 5 stars. These are the stars that make up the head of the Whale. The best time to see the Whale is October through January.

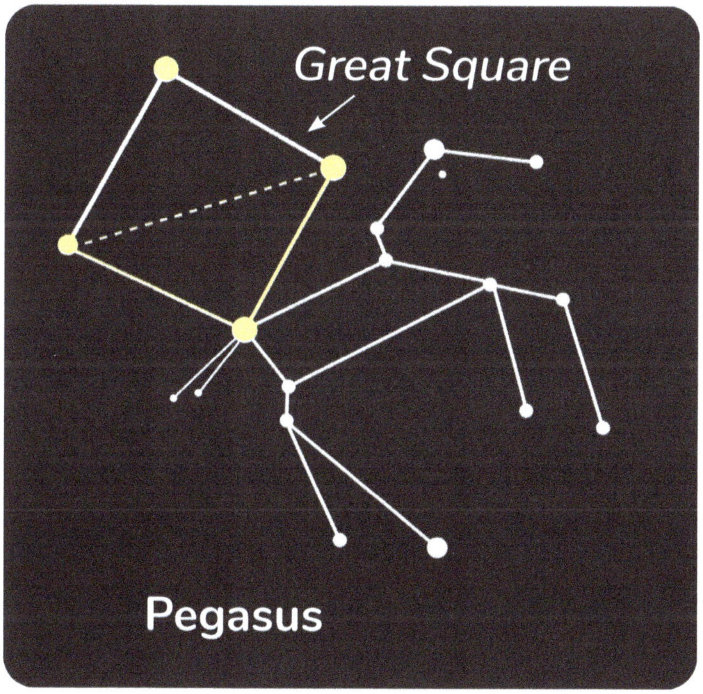

Another favorite constellation in the Southern Hemisphere is **Pegasus**, the winged horse. Part of the Pegasus constellation includes three of the four stars of the Great Square. These three stars make up the wing of Pegasus which sits on the hind end of the horse. Two little stars near the hind end make up the tail and the head extends in the opposite direction from the tail. The best time of the year to see Pegasus is from August to October.

19.4 Using Stars to Navigate

How do you find your way to the grocery store? How do you know which street to take to the park? If you need to go to a friend's house, do you turn to your right, left, or go straight ahead from your front door?

In each of these situations, you are navigating your way from one place to another place. **Navigation** simply means to make one's way from one location to another. There are several different techniques people use to navigate.

▶ One way to navigate is to use **landmarks**.

If you walk to the grocery store with your parents, you might notice that it is located just across from a park and next to a gas station. The next time you need to go to the grocery store, you can use your knowledge of the park and the gas station as landmarks to guide you to the store. However, if you were far out to sea where there are no landmarks, how would you find your way home?

This was a problem for early sailors. When traveling along the coast, they could use landmarks to find their way, but what happened when they traveled far enough out in the ocean that they could no longer see the shore? Early sailors discovered that they could use the stars as a way to navigate across the sea. Using the stars is a great way to find your way, whether you are on land, sea, in an airplane or even in space!

One easy star to use for navigation in the Northern Hemisphere is Polaris. Unlike other stars, Polaris actually stays in the same place in the sky and doesn't appear to move. The north pole of Earth's axis points almost directly at Polaris. If you can find Polaris, you can tell which way is north, and once you locate north, you can find south, east, and west.

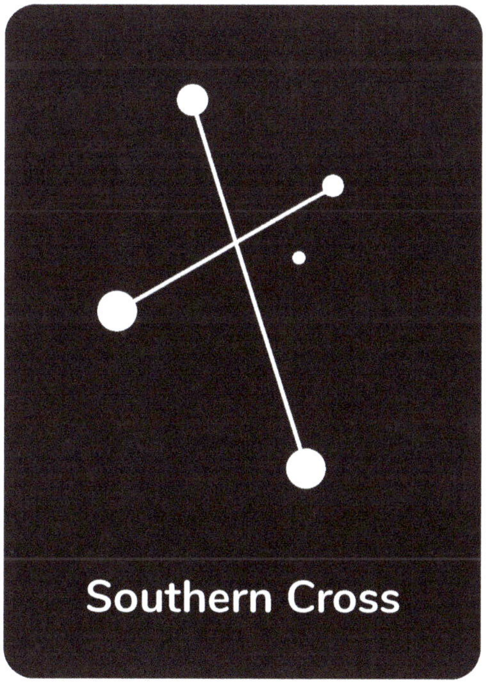

In the Southern Hemisphere people use the **Southern Cross** constellation for navigation. Although Earth's axis at the South Pole doesn't point directly at an individual star, the two stars that form the long part of the constellation can be used to find south.

19.5 Summary

- A **constellation** is a group of stars that together form a particular shape.
- Common **constellations** in the Northern Hemisphere include the **Big Dipper**, the **Great Bear**, and **Orion**.
- Common **constellations** in the Southern Hemisphere include the **Whale** and **Pegasus**, both of which can be located by finding the **Great Square**.
- Stars can be used to **navigate**, or find the way, from one location to another.

19.6 Vocabulary

Basic Words

whale (WAYL)

Challenge Words

Cassiopeia (kaa-see-uh-PEE-uh)

constellation (kahn-stuh-LAY-shuhn)

navigation (naa-vuh-GAY-shuhn)

hemisphere (HEH-muh-sfihr)

Orion (uh-RIY-uhn)

Pegasus (PEH-guh-suhs)

Polaris (puh-LEHR-uhs)

CHAPTER 20

Earth's NEIGHBORHOOD

20.1 Introduction

In our solar system there are two different types of planets. Some of our planetary neighbors are Earth-like, or **terrestrial**, and some are Jupiter-like, or **Jovian**. In this chapter we will take a look at where our planetary neighbors "live" in our solar neighborhood.

20.2 Our Solar Neighborhood

Most people live in some kind of neighborhood. A neighborhood is an area of town with houses, apartments, a few businesses, and possibly a park.

If you take a walk down the block in your neighborhood, you can see where your neighbors live. Some of your neighbors live close to you. Maybe they live next door and share the same backyard. Other neighbors live farther away, but they may all go to the local grocery store or walk their dog in the local park. We would say that all of the people who live in this particular area of town are part of a neighborhood.

In the same way, planets share a particular area in space.

A **solar system** is made up of a sun and the planets and other objects that travel around that sun.

In our solar system, there are eight planets. All of the planets share the same sun. Some planets are closer to our Sun, and some are farther away, just like the neighbors in your neighborhood.

The closest planet to the Sun and the smallest planet is **Mercury**. Because Mercury is so close to the Sun, its surface can be very hot. The temperature at noon on Mercury can get up to as much as 425 degrees Celsius (800 degrees Fahrenheit)! But Mercury does not have enough air to hold onto the heat from the Sun. At night the temperature on Mercury can go down to below 18 degrees Celsius (below zero degrees Fahrenheit). So Mercury does not have the right temperatures for plants and animals to be able to live.

The next closet planet to the Sun is **Venus**. Venus is about twice as far away from the Sun as Mercury. However, even though Venus is farther away from the

Sun than Mercury is, Venus is actually hotter! Venus has lots of carbon dioxide in the air. This heats up the surface and holds the heat so Venus is hot all the time. The surface of Venus can reach over 460 degrees Celsius (860 degrees Fahrenheit). Venus is much too hot to support plant and animal life.

The next closest planet to the Sun is **Earth**. Earth is close enough to the Sun to have enough heat for life to exist, but not so close that it is too hot for living things. Earth is the only planet in our solar system that supports plant and animal life.

Mars sits farther away from the Sun than Earth does. Mars is much colder than Earth because it is farther away from the Sun. However, Mars is almost close enough to the Sun to support life.

Mercury, Venus, Earth, and Mars make up the **inner solar system**, or inner neighborhood. From an astronomer's perspective, all these planets are relatively close to each other.

Much farther out are the four planets in the **outer solar system**. **Jupiter** is the first planet in the outer solar system. Jupiter is more than five times farther away from the Sun than Earth is.

Saturn is even farther away from the Sun than Jupiter is. Saturn is the second planet in the outer solar system.

Uranus and **Neptune** are the last two planets in the outer solar system.

Neptune is almost 30 times as far away from the Sun as Earth is. Jupiter, Saturn, Uranus, and Neptune are all much too cold to support plant and animal life.

20.3 Orbits

The planets don't just sit in one spot, but move in a nearly circular **orbit** around the Sun. An orbit is a particular path, like a road, that a planet follows.

Each planet stays in its orbit at its particular distance away from the Sun. Planets don't cross other planetary orbits or ever bump into each other.

Each planet takes a certain number of days to orbit the Sun. This is called a **planetary year**. Mercury orbits the Sun faster than does any other planet. It only takes 88 days for Mercury to complete one orbit. So one year on Mercury is only 88 days.

It takes Venus a little longer than Mercury to orbit the Sun, but not as long as it takes Earth. Venus orbits the Sun in 225 days, and Earth orbits the Sun in 365 days. The length of our calendar year is 365 days.

It takes Mars 687 days to orbit the Sun, and it takes Jupiter almost 12 Earth years to complete one planetary year.

Saturn takes almost 30 Earth years to orbit the Sun, and Uranus orbits the Sun in 84 Earth years. If you lived on Neptune you would need 165 Earth years just to get around the Sun once!

Planet	Number of Days for One Orbit of the Sun
 Mercury	88
 Venus	225
 Earth	365
 Mars	687
 Jupiter	4,332
 Saturn	10,760
 Uranus	30,700
 Neptune	60,200

20.4 Why Is Earth Special?

Of all the planets in our solar system, only Earth is the right distance away from the Sun, with the right combination of water, oxygen, minerals, and soils to support plant and animal life. Earth has just the right conditions for plant and animal life to exist.

Earth has many unique features that make it just right for plant and animal life. If Earth were closer to the Sun, it would be too hot for life. If Earth were farther away from the Sun, it would be too cold for life. If Earth's atmosphere had too much gas, like Venus does, it would cause the Earth's surface to heat up and would make it too hot.

Earth is special in this way. There is no other planet in our solar system that can support plant and animal life. And so far, no other planet in the universe has been found that supports plant and animal life. Earth has just the right temperature, is just the right distance from the Sun, and is made of just the right materials for plant and animal life as we know it to exist.

20.5 Summary

- We live together with other planets in a planetary neighborhood called the **solar system**.
- All of the planets in our **solar system** share the same sun.
- **Mercury** is closest to the Sun, followed by **Venus**, **Earth**, **Mars**, **Jupiter**, **Saturn**, **Uranus**, and **Neptune**.
- Each planet rotates around the Sun in an **orbit**.
- Each planet takes a different number of days to complete one orbit around the Sun. The number of days it takes a planet to orbit the Sun once is called a **planetary year**. Our **planetary year** is 365 days.

20.6 Vocabulary

Basic Words

Earth (UHRTH)

Mars (MAHRZ)

Neptune (NEHP-toon)

Saturn (SAA-tuhrn)

Venus (VEE-nuhs)

Challenge Words

Jovian (JOH-vee-uhn)

Jupiter (JOO-puh-tuhr)

Mercury (MUHR-kyuh-ree)

planetary (PLAA-nuh-tehr-ee)

terrestrial (tuh-REH-stree-uhl)

Uranus (YOOR-uh-nuhs)

CHAPTER 21

Beyond the NEIGHBORHOOD

21.1 Introduction

In Chapter 20 we looked at our planetary neighborhood called the solar system. We saw that eight planets orbit our Sun and why Earth is the only planet in our solar system that can support plant and animal life as we know it.

But what about other solar systems? What lies beyond our neighborhood? Are there other neighborhoods with suns like ours and planets like ours that support plant and animal life?

21.2 Nearest Star

If you look up into the sky on a dark, clear night, you can see lots of stars. On a moonless night, far away from the city lights, you might be able to see 2,000 stars or more.

The closest star to our Earth is a star called **Proxima Centauri**.

Proxima Centauri is actually part of a three-star system. The two other stars in this system are called **Alpha Centauri A** and **Alpha Centauri B.** Alpha Centauri A, Alpha Centauri B, and Proxima Centauri orbit each other.

Proxima Centauri is smaller and cooler than our Sun and is red in color. Even though Proxima Centauri is the closest star to our Sun, it is not bright enough to be seen without a telescope.

21.3 Brightest Star

The brightest star we can see in the sky is called **Sirius**. Sirius is farther away than Proxima Centauri but is brighter. The brightness of a star does not depend on how close it is, but on how much light the star makes.

One way to think about how much light a star makes is to look at the difference between a candle and a flashlight. A candle will light up or **illuminate** an area of a meter or so (a few feet) around it, but a flashlight can illuminate several meters (yards). A flashlight puts out more energy than a candle and as a result can illuminate a much longer distance.

If you were to look at a candle and a flashlight from several meters (yards) away, the flashlight would look brighter. If you moved the flashlight a few meters (yards) farther away than the candle, it would still look brighter.

In the same way, Sirius is a brighter star than Proxima Centauri even though it is farther away.

21.4 Biggest Star

The biggest star is not necessarily the closest star or the brightest star. The biggest star we can see from Earth is called **VY Canis Majoris** or **VY CMa**.

VY CMa is about 2,000 times larger than our Sun. If VY CMa were in our solar system replacing our Sun, it would extend beyond the orbit of Saturn!

21.5 Stars With Planets

Scientists think that most stars have planets orbiting them.

It is very difficult for astronomers to see the planets in other solar systems directly. Planets are very dim because they don't produce their own light like suns do. However, astronomers can use other techniques to figure out if stars have planets orbiting them. One technique is to use physics to determine if a star is wobbling. As planets orbit a sun, their gravity pulls on the sun, causing it to wobble. Astronomers can study a sun's wobble to determine if one or more planets might exist in orbit around the sun.

Many planets in other solar systems have been found, but so far none of them appear to be like Earth. In order for a planet to be able support plant and animal life, the planet would need to be close to a sun. However, if it were too close, the planet would be too hot. If the planet were too far away from the sun, it would be too cold. Also, the sun would have to be large enough to produce enough energy for plants and animals to use.

Scientists continue to search for life on other planets. So far, Earth is the only planet that we know supports plant and animal life.

21.6 Summary

- Our nearest neighboring star is **Proxima Centauri**.
- The brightest star is **Sirius**.
- The largest star is **VY Canis Majoris (VY CMa)**.
- So far, Earth is the only planet we know of that supports plant and animal life.

21.7 Vocabulary

Challenge Words

Alpha Centauri A (AAL-fuh sehn-TAWR-ee AY)

Alpha Centauri B (AAL-fuh sehn-TAWR-ee BEE)

illuminate (ih-LOO-muh-nayt)

Proxima Centauri (PRAHK-suh-muh sehn-TAWR-ee)

Sirius (SIHR-ee-uhs)

VY Canis Majoris (VEE WIY KAY-nuhs muh-JAWR-uhs)

VY CMa (VEE WIY SEE EHM AY)

CHAPTER 22

Putting It All TOGETHER

22.1 Tools and Science

In this book you learned about the many tools scientists use to study the world around us. The tools scientists use vary depending on what is being studied. Also, in each of the **core subjects** of **chemistry**, **biology**, **physics**, **geology**, and **astronomy**, there are a few **basic tools** and a greater number of **advanced tools**.

22.2 Sharing Tools

Scientists often share their tools with other scientists who are studying different core subjects. For example, a chemist may use some of the same tools as a biologist, and a biologist may use some of the same tools as a chemist. For instance, while performing an experiment to study small molecules in a cell, a biologist might need to use a beaker, a basic tool commonly used by chemists.

A chemist studying the atoms in a geode might need to use a geologist's rock hammer to break the geode open. Or an astronomer looking for life on other objects in space might use a biologist's microscope to examine moon rocks to look for tiny archaea or bacteria.

22.3 Working Together

Most scientists specialize in a specific subject area, but in many research labs scientists who specialize in different subjects work together. It is not uncommon to see chemists working side-by-side with biologists and physicists to solve a specific scientific problem.

- Teams of scientists working together to solve problems is a great way to perform experiments.

Many problems need to be examined from different viewpoints in order to find a solution. Chemists may think of solving problems in ways that are different from biologists. Also, because there is so much scientific information that has been discovered and studied, no one scientist can learn everything there is to know about science. Therefore, having a group of scientists **collaborate** (work together) on a project is the best way to learn.

22.4 Summary

- Scientists who study **chemistry**, **biology**, **physics**, **geology**, and **astronomy** use both **basic** and **advanced tools**.
- Scientists who specialize in a **core subject** learn how to use **tools** specific to their subject.
- Scientists share **tools** with each other.
- Scientists from different **core subject** areas work in groups and **collaborate** on projects.

22.5 Vocabulary

Review Words

astronomy (uh-STRAH-nuh-mee)

biology (biy-AH-luh-jee)

chemistry (KEH-muh-stree)

geology (jee-AH-luh-jee)

physics (FIH-zihks)

Challenge Words

collaborate (kuh-LAA-buh-rayt)

Glossary / Index

(Pronunciation Key at end)

absorb (uhb-ZAWRB) • to take in, 54, 113

acceleration (ihk-seh-luh-RAY-shuhn) • the change in speed of a moving object for a given time, 77-78

acid (AA-suhd) • a type of molecule that gives foods a sour taste, 16-38

acid-base indicator • see indicator, acid-base

acid-base reaction (AA-suhd BAYSS ree-AAK-shuhn) • a special kind of chemical reaction between an acid and a base, 16-30

advanced telescope • see telescope, advanced

advanced tool • see tool, advanced

air (AYR) • the substance we breathe, 95, 97, 110-115

alchemist (AAL-kuh-mihst) • an early chemist who learned how different chemicals react, 10

alembic (uh-LEHM-bihk) • an early chemistry tool that is still used, 10

Alpha Centauri A (AAL-fuh sen-TAWR-ee AY) • a star in the Alpha Centauri system, 144

Alpha Centauri B (AAL-fuh sen-TAWR-ee BEE) • a star in the Alpha Centauri system, 144

amoeba (uh-MEE-buh) • a protist that moves by using pseudopods, 48, 53-54, 55

amoeboid (uh-MEE-boid) • a protist that moves by using pseudopods, 48, 49

analytical scale • see scale, analytical

Archer (AHR-chuhr) • a constellation in the Southern Hemisphere, 125

argon (AHR-gahn) • an element found as a gas in air, 112, 115

Aristotle (AA-ruh-stah-tuhl) • [384-322 B.C.E.] Greek philosopher who organized plants and animals into categories, 40-41, 68

asthenosphere (aas-THEH-nuh-sfeer) • the soft, putty-like layer of the mantle that lies below the lithosphere, 103-108

astronomy (uh-STRAH-nuh-mee) • the study of the cosmos, 117-148

atmosphere (AAT-muh-sfeer) • the air layer of Earth, 95-97, 99-101, 111

atom (AA-tuhm) • the smallest unit of matter, 16, 18, 26-30, 33-35, 112-113

average speed • see speed, average

balance (BAA-lentz) • a tool used to measure weight, 12, 63

base (BAYSS) • a type of molecule that can make things feel slippery or have a bitter taste, 16-38

basic tool • see tool, basic

beaker (BEE-kuhr) • a cup shaped tool used in science laboratories to measure and hold liquids, 5, 11-12

Big Dipper • a constellation in the Northern Hemisphere, 125-127

biologist (biy-AH-luh-jihst) • a scientist who studies living things, 39-43

biologist, field • see field biologist

biology (biy-AH-luh-jee) • the study of life, 39-60

biome (BIY-ohm) • a particular part of the biosphere in which the climate, soils, and plant life are similar throughout, 97

biosphere (BIY-uh-sfihr) • the part of Earth where living things exist, 95-97, 99, 101

Buchner flask (BUHK-nuhr FLAASK) • a type of glassware used in a chemistry lab, 13

budding (BUHD-ing) • the process of cell division in yeast, 59-60

carbohydrate (kahr-boh-HIY-drayt) • a long chain of sugar molecules, 36-37

carbon dioxide (KAHR-buhn diy-AHK-siyd) • a molecule made of one carbon atom and 2 oxygen atoms; a gas found in the air, 59, 95, 112-115, 136

Cassiopeia (kaa-see-uh-PEE-uh) • a constellation in the Northern Hemisphere, 125

chemical reaction (KEH-muh-kuhl ree-AAK-shuhn) • occurs when atoms and molecules meet and trade places, join together, or separate from each other; also called a reaction, 16, 20, 22, 26-30

chemistry (KEH-muh-stree) • the investigation of what physical things are made of and the ways in which they change; the study of matter, 9-38

chisel edge rock hammer • see hammer, chisel edge rock

chlorophyll (KLAW-ruh-fihl) • a molecule in living things that captures sunlight and uses it to make food, 51

chloroplast (KLAW-ruh-plaast) • an organelle that contains chlorophyll; used by some organisms to make food, 51, 55

chromatograph • see gas chromatograph

ciliate (SIH-lee-uht) • a protist that moves by using cilia, 48, 49

cilium (SIH-lee-uhm) [*plural,* **cilia** (SIH-lee-uh)] • a small hairlike feature on the body of some protists, 48, 49, 52-53

Coleps (KOH-leps) • a protist that rotates its whole body to swim through the water and to eat, 54

collaborate (kuh-LAA-buh-rayt) • to work together with others, 151

computer (kuhm-PYOO-tuhr) • an electronic device that can store, retrieve, and process data, 12, 65, 91-92, 117

constellation (kahn-stuh-LAY-shuhn) • a group of stars that seems to form a shape or image in the sky, 124-132

Copernicus, Nicolaus (koh-PUHR-ni-kuhss, NIH-koh-lahs) • (1473-1543 C.E.) Polish astronomer; used mathematics to show that the Earth moves around the Sun, 118

core (KAWR) • the very center part of Earth, 102-105, 108

core, inner (KAWR, IH-nuhr) • the solid inner part of Earth's core, 103-105

core, outer (KAWR, OW-tuhr)• the liquid rock and metal outer layer of Earth's core, 103-105, 108

core subject (KAWR SUHB-jehct) • one of the major areas of scientific study: chemistry, biology, physics, geology, astronomy, 149

crack hammer • see hammer, crack

crust (KRUHST) • in geology, the outer layer of Earth, 94, 102-108

detail (DEE-tayl) • a small, particular part of a thing, 63

dynamic testing machine (diy-NAA-mihk TEHST-ihng muh-SHEEN) • an advanced tool used to test the properties of objects and materials, 64

Earth (UHRTH) • the planet we live on, 134, 136, 138-141

educated guess (EH-juh-kay-tuhd GESS) • a guess or idea based on scientific data, 105-107

Erlenmeyer (UHR-luhn-miy-uhr) **bulb** • a type of glassware used in a chemistry lab, 13

euglena (yoo-GLEE-nuh) • a protist that moves using a flagellum and uses photosynthesis to make food, 47, 51,55

extend (ihk-STEHND) • to stretch out, 5

eyepiece (IY-peess) • at one end of a telescope, the lens that the eye looks through, 119, 121

eyespot (IY-spaht) • an area in a euglena that can detect light, 51

false feet (FAWLSS FEET) • pseudopods, 48, 54

field biologist (FEELD biy-AH-luh-jihst) • a scientist who studies living things outdoors, 41

field notebook (FEELD NOHT-buuk) • a place for a field biologist to record facts such as the date, time of day, location, and the plants or animals being observed, 41

five senses (FIYV SEHN-suhs) • sight, hearing, smell, touch, and taste, 4, 7

flagellate (FLAA-juh-layt) • a protist that moves by using a flagellum, 47, 49

flagellum (fluh-JEH-luhm) [*plural,* **flagella** (fluh-JEH-luh)] • a whiplike structure on some types of cells and organisms, 47

flask (FLAASK) • a type of glass container used in labs that has a narrowed neck, 13

force (FAWRSS) • any action that changes the location of an object, the shape of an object, or how fast or slowly an object is moving, 68-69, 71-72

friction (FRIHK-shuhn) • a force that causes moving objects to slow down and stop, 71-72

Fungi (FUHN-jiy) or (FUHN-giy) • in taxonomy, the kingdom that includes mushrooms,molds, and yeasts, 57

fungus (FUHN-guhs) [*plural,* **fungi** (FUN-jiy) or (FUHN-giy)] • a member of the kingdom Fungi, 56-60

galaxy (GAA-luhk-see) • a large group of stars, gas, dust, planets, and other objects in space, 121

Galileo Galilei (gaa-luh-LAY-oh gaa-luh-LAY)] • [1564-1642 C.E.] Italian physicist and astronomer; studied motion, 68, 69, 118, 122

gas chromatograph (GAAS kroh-MAA-tuh-graaf) • an instrument that can detect different types of gases in very small amounts, 12-13

gas molecule (GAAS MAH-luh-kyool) • two or more atoms of a gas joined together, 112, 113

gear (GEER) • a simple machine that uses rotational motion, 83-86

geochemist (jee-oh-KEH-mihst) • a geologist who studies the chemistry of Earth, 99

geographic information system (GIS) (jee-uh-GRAA-fihk ihn-fawr-MAY-shuhn SIHS-tuhm) (JEE IY ESS) • a computer system that is used to gather and analyze geographical data, 92

geologist (jee-AH-luh-jihst) • a scientist who studies Earth, 87-92

geologist, structural (jee-AH-luh-jihst, STRUHK-chuh-ruhl) • a geologist who studies how Earth is put together and how it changes, 99

geology (jee-AH-luh-jee) • the study of Earth, 87-115

geosphere (JEE-oh-sfihr) • Earth's crust, mantle, and core together; made of rock, minerals, and soils, 94, 96, 97, 99, 102-108

GIS (JEE IY ESS) • geographic information system; a computer system that is used to gather and analyze geographical data, 92

glassware (GLAAS-wayr) • tools made of glass that are used in a science lab, 13

global positioning system (GPS) (GLOH-buhl puh-ZIH-shuhn-ihng SIHS-tuhm) (JEE PEE ESS) • an electronic device that uses signals from satellites to determine position on Earth, 91

GPS (JEE PEE ESS) • global positioning system, 91

gravity (GRAA-vuh-tee) • the force that pulls everything on or near Earth toward the center of Earth, 115

Great Bear (GRAYT BAYR) • a constellation in the Northern Hemisphere, 127

Great Square (GRAYT SKWAYR) • a constellation in the Southern Hemisphere, 128, 129

Gregorian telescope • see telescope, Gregorian

hammer (HAA-muhr) • a T-shaped tool used for pounding, 5, 87-90

hammer, chisel edge rock (HAA-muhr, CHIH-zuhl EHJ RAHK) • a hammer that has a head with one flat, square end and one broad, chisel-like end; used to break rocks, 89

hammer, crack (HAA-muhr, KRAAK) • a heavy hammer with two blunt ends; used to break rocks, 89-90

hammer, pointed tip rock (HAA-muhr, POIN-tuhd tihp RAHK) • a hammer with a head that is flat at one end and pointed at the other; used to break rocks; also called a rock pick, 89-90

hammer, rock (HAA-muhr, RAHK) • a hammer used by geologists to break rocks, 89-90

hand lens (HAAND LEHNZ) • a small handheld magnifier that folds into a holder, 90

H group (AYCH GROOP) • a hydrogen atom in an acid molecule, 18-19, 24, 27

Hubble Space Telescope (HUH-buhl SPAYSS TEH-luh-skohp) • a telescope that is in orbit around the Earth outside the atmosphere, 121

hydrogen chloride (HIY-druh-juhn KLAW-riyd) • an acid, 19, 26

hydrosphere (HIY-droh-sfihr) • the part of Earth where water exists, 95-97, 99, 101

hypha (HIY-fuh) [*plural,* **hyphae** (HIY-fee) • a thin, thread-like structure in a fungus; used to break down dead organisms for food, 58, 60

Ibn al-Haytham (IBN aal-HAY-thuhm) • [965-1040 C.E.] Muslim philosopher who studied light and how light rays travel, 62

illuminate (ih-LOO-muh-nayt) • to light up, 145

indicator (IHN-duh-kay-tuhr) • something that tells us something is happening or has happened, 22-23, 36, 38

indicator, acid-base (IHN-duh-kay-tuhr, AA-suhd BAYSS) • something that tells us whether a liquid is an acid or a base, 23, 36

inertia (ih-NUHR-shuh) • the tendency of an object to resist a change in motion, 69-72

inner core • see core, inner

inner solar system • see solar system, inner

instrument (IHN-struh-muhnt) • a more complicated tool that is used to do detailed and exact work, 1, 4, 6, 7, 11, 12-14, 42-43, 64, 66, 91-92, 119

invention (ihn-VEHN-shuhn) • something new that is created to help people do a certain task, 2

irregular galaxy • see galaxy, irregular

Jovian (JOH-vee-uhn) • Jupiter-like, 134

Jupiter (JOO-puh-tuhr) • a large, gaseous planet in the Earth's solar system, 134, 136-137, 139

kinetic energy • see energy, kinetic

kingdom (KIHNG-duhm) • a group in taxonomy that is a subdivision of a domain, 47, 57

kit (KIHT) • a group of tools or objects used for a certain purpose, 42

landmark (LAAND-mahrk) • an object that marks a location and can help with navigation, 130-131

Leeuwenhoek, Anton van (LAY-vuhn-huhk, AAN-tahn vahn) • [1632-1723 C.E.] lens maker from Holland; first person to see protists, 46

lens (LEHNZ) • a piece of clear glass or plastic often used to make objects look larger, 5, 43, 90, 119-121

linear motion • see motion, linear

Lippershey, Hans (LIH-puhr-shee, HAHNZ) • (1570-1619 C.E.) a Dutch lens maker; invented the spyglass, 118

liquid (LIH-kwihd) • any pourable substance, such as water, 11-12

lithosphere (LIH-thuh-sfihr) • the hard, rocky layer of the mantle just below the crust, 103-107

magma (MAAG-muh) • molten rock found deep inside Earth, 104, 106

magnetic field (maag-NEH-tihk FEELD) • the area of force that surrounds a magnet; in geology, the magnetic force that begins within the core and surrounds Earth, 95, 108

magnetosphere (maag-NEE-tuh-sfihr) • the area in space surrounding Earth that contains the magnetic field, 95-97, 99, 101, 108

magnify (MAAG-nuh-fiy) • to make an object look larger, 5, 42-43, 90, 119

magnifying glass (MAAG-nuh-fiy-ing GLAASS) • a basic tool used to make objects look larger, 42-43

magnifying lens (MAAG-nuh-fiy-ihng LEHNZ) • a piece of clear glass or plastic used to make objects look larger, 5

mantle (MAAN-tuhl) • in geology, the layer of Earth below the crust, 102-105

mantle, lower (MAAN-tuhl, LOH-uhr) • the mesosphere, 105

mantle, upper (MAAN-tuhl, UH-puhr)• the lithosphere and the asthenosphere grouped together, 105

Mars (MAHRZ) • a planet in our solar system; the 4th from the Sun, 136, 139

mass (MAASS) • in simple terms, how heavy an object is, 70

measure (MEH-zhuhr) • to determine the amount of something, 1, 6-7, 11-12, 39, 61-66, 75-76

Mercury (MUHR-kyuh-ree) • a planet in our solar system; the closest planet to the Sun, 135-136, 138, 139

mesosphere (MEH-zoh-sfihr) • the deepest layer of the mantle, 103-105, 108

microscope (MIY-kruh-skohp) • an instrument that magnifies things that are too small to be seen with the eyes alone, 7, 43, 46-47

microscope, scanning tunneling (MIY-kruh-skohp, SKAAN-ing TUH-nuhl-ing) • an instrument used to see atoms, 7

mold (MOHLD) • a type of small fungus, 57, 60

molecule (MAH-luh-kyool) • two or more atoms joined together, 16-38

monitor (MAH-nuh-tuhr) • to watch or keep track of, 39

motion, linear (MOH-shuhn, LIH-nee-uhr) • motion that follows a straight line, 74-78, 81, 83

motion, nonlinear (MOH-shuhn, nahn-LIH-nee-uhr) • motion that occurs along a curved path rather than a straight line, 80-86

motion, rotational (MOH-shuhn, roh-TAY-shuh-nuhl) • a type of non-linear motion that occurs around a center point, 82-83, 86

mushroom (MUHSH-room) • a type of fungus, 56-60

navigation (naa-vuh-GAY-shuhn) • making one's way from one location to another, 127, 130-132

Neptune (NEHP-toon) • a Jovian planet in the Earth's solar system, eighth from the Sun, 137, 139

Newtonian telescope • see telescope, Newtonian

nitrogen (NIY-truh-juhn) • an element (atom), 112, 115

nonlinear motion • see motion, nonlinear

North Star (NAWRTH STAHR) • the star that is above the northernmost point of the Earth; can be used as a navigational marker; also called Polaris, 127

objective lens (uhb-JEHK-tihv LEHNZ) • the lens of a telescope that collects light so an object can be viewed, 119, 121

observation (ahb-suhr-VAY-shuhn) • the act of taking note of a fact or occurrence, 4, 41, 44

observe (uhb-ZUHRV) • to take note of, 4-7, 39-44, 46

OH group (OH AYCH GROOP) • an oxygen atom and a hydrogen atom in a base molecule, 18-19, 24, 27

Omar Khayyam (OH-mahr kiy-YAAM) • [1048-1131 C.E.] a Persian scientist who accurately calculated the length of a year, 63

orbit (AWR-buht) • the curved path of a celestial body as it travels around another celestial body, 137-139, 141, 147

Orion the Hunter (uh-RIY-uhn thuh HUHN-tuhr) • a constellation in the Northern Hemisphere, 128

oscilloscope (ah-SIHL-uh-skohp) • a tool used to measure electrical waves, 65

outer core • see core, outer

outer solar system • see solar system, outer

oxygen (AHK-sih-juhn) • an element; a gas found in the air that we need in order to live, 18-19, 112-115

paramecium (pehr-uh-MEE-shee-uhm) [*plural*, **paramecia** (per-uh-MEE-shee-uh) • a type of protist that has cilia, 48, 52-53, 55

Pegasus (PEH-guh-suhs) • a constellation in the Southern Hemisphere, 129

pelican (PEH-lih-kuhn) • in science, an early chemistry tool that is still used, 10-11

physics (FIH-zihks) • the branch of science that explores how the natural world works, 61-86

planetary year (PLAA-nuh-tuh-ree YEER) • the number of days it takes a planet to orbit the Sun, 138-139

plate (PLAYT) • in geology, a huge piece of the hard, outer mantle, 104, 106

pointed tip rock hammer • see hammer, pointed tip rock

Polaris (puh-LEHR-uhs) • the star that is above the northernmost point of Earth, can be used as a navigational marker; also called the North Star, 126, 127, 131

pollute (puh-LOOT) • to make air, water, or land dirty, 111

pollution (puh-LOO-shuhn) • substances in the air, water, or land that make it dirty and unsafe to use or breathe, 111

protist (PROH-tihst) • a tiny creature too small to see without a microscope; in the kingdom Protista; also called a protozoan, 45-55

Protista (proh-TIH-stuh) • in taxonomy, the kingdom that contains protists, 47

protozoan (proh-tuh-ZOH-uhn) [*plural,* **protozoa** (proh-tuh-ZOH-uh)] • a protist, 46, 48

Proxima Centauri (PRAHK-suh-muh sen-TAW-ree) • the star that is nearest to our solar system; part of the Alpha Centauri system, 144, 145

pseudopod (SOO-duh-pahd) • in an amoeba, a false foot used for crawling and eating, 48, 49, 54, 55

radio (RAY-dee-oh) • a device that sends and receives signals, 91

reaction • see chemical reaction

relative (REH-luh-tiv) • in comparison to something else, 75

rock hammer • see hammer, rock

rock pick • see pointed tip rock hammer

rotate (ROH-tayt) • to turn around a center point, 54, 86, 141

rotational motion • see motion, rotational

ruler (ROO-luhr) • a basic tool used for making measurements, 63

salt • one of the products of an acid-base reaction, (see also table salt), 29, 30, 34

sample (SAM-puhl) • a thing or part of a thing being studied, 43

satellite (SAA-tuh-liyt) • a machine that is put into orbit around the Earth or another celestial body, 91, 117

Saturn (SAA-tuhrn) • a Jovian planet in Earth's solar system, sixth from the Sun; has rings, 137, 139

scale (SKAYL) • a tool used to measure weight, 6, 12

scale, analytical (SKAYL aa-nuh-LIH-tuh-kuhl) • a tool used to accurately measure the weight of a small amount of a material, 6

scanning tunneling microscope • see microscope, scanning tunneling

sense (SEHNSS) • a specialized mechanism by which an animal receives and responds to stimuli (for example, sight); to perceive or become aware of, 4, 7, 35, 46

shark cage (SHAHRK KAYJ) • a protective cage from which a diver can safely observe sharks underwater, 43

Shen Kuo (SHEHN GWOH) • [1031-1095 C.E.] Chinese philosopher; described the magnetic needle compass, 63

Sirius (SIHR-ee-uhss) • the brightest star in the sky, 145

sodium chloride (SOH-dee-uhm KLAW-riyd) • a molecule made of a sodium atom joined to a chlorine atom; table salt, 29, 34

sodium hydroxide (SOH-dee-uhm hiy-DRAHK-siyd) • a base, 19, 26

solar system (SOH-luhr SIHS-tuhm) • a star and the objects that orbit it, 121, 135-137, 141

solar system, inner (SOH-luhr SIHS-tuhm, IH-nuhr) • the four planets closest to the Sun—Mercury, Venus, Earth, and Mars, 136

solar system, outer (SOH-luhr SIHS-tuhm, OW-tuhr) • the four planets farthest from the Sun—Jupiter, Saturn, Uranus, and Neptune, 136-137

sound meter (SOWND MEE-tuhr) • a tool used to measure high and low sounds, 64

Southern Cross (SUH-thuhrn KRAWSS) • a constellation in the Southern Hemisphere that can be used for navigation, 131

spaceship (SPAYSS-shihp) • a vehicle that can travel beyond Earth, 117

specialized (SPEH-shuh-liyzd) • having a particular purpose, 13

speed • how fast an object moves, 75-78

speed, average (AA-vuh-rihj) • a value arrived at by adding together a set of measurements of speed and then dividing by how many measurements there are, 76-77, 78

sphere (SFIHR) • a ball-shaped object; in geology, one of several parts of the Earth that work together, 94-101

spore (SPAWR) • the reproductive unit of molds, mushrooms, and some other organisms, 57, 59, 60

sprocket (SPRAH-kuht) • a wheel that has teeth on it and is used in a machine, 83-86

spyglass (SPIY-glaass) • a tube with a lens at each end that is used to make stars look bigger, 118, 122

stage (STAYJ) • the part of a microscope where the sample is placed, 43

structural geologist • see geologist, structural

submarine (SUHB-muh-reen) • an underwater boat, 43

table salt (TAY-buhl SAWLT) • a molecule made of a sodium atom joined to a chlorine atom; sodium chloride, 29, 34

technology (tehk-NAH-luh-jee) • a machine or piece of equipment developed by using scientific knowledge to solve a problem, 91-92

telescope (TEH-luh-skohp) • a tool or instrument that helps astronomers see far into the distance; makes faraway objects look larger, 117-122

telescope, advanced (TEH-luh-skohp, aad-VANST) • a more complicated telescope that is able to see objects farther away in space, 121

telescope, Gregorian (TEH-luh-skohp, grih-GAW-ree-uhn) • a telescope that has mirrors on the inside of the tube, 120-121

telescope, Newtonian (TEH-luh-skohp, noo-TOH-nee-uhn) • a telescope that has mirrors on the inside of the tube, 120-121

terrestrial (tuh-REHS-tree-uhl) • Earth-like, 134

Thales of Miletus (THAY-leez uhv miy-LEE-tuhs) • (625–545 B.C.E.) Greek philosopher who learned about geometry and astronomy in Egypt, 62

Theophrastus (thee-uh-FRAA-stuhs) • (372–287 B.C.E.) Greek philosopher who explored the nature of stones, 88

thermal imager (THUHR-muhl IH-mih-juhr) • a tool used to measure small changes in temperature, 64

thermometer (thuhr-MAH muh-tuhr) • an instrument used to measure temperature, 5, 22, 63

timer (TIY-muhr) • a basic tool used to keep track of time, 63

tongue (TUHNG) • a muscle in the mouth that detects certain atoms and molecules, 34-38

tool • a device used to help in performing a task, 1-14, 22, 39, 41-44, 61-66, 87-92, 117-122, 149-151

tool, advanced (aad-VAANST) • a more complicated tool used to do detailed and exact work; an instrument, 1, 6-7, 11-14, 42, 65, 66, 91-92, 121, 149

tool, basic (BAY-sihk) • a simple handheld tool, 1, 4-5, 7, 11-12, 14, 41-42, 63, 66, 89-90, 92, 117-118, 149

track (TRAAK) • to observe the path of a thing that is moving, 39, 43

tracking equipment (TRAAK-ihng ih-KWIHP-muhnt) • tools used for observing the movement of an object, 43

Uranus (YOOR-uh-nuhs) • a Jovian planet, seventh from the Sun in the Earth's solar system, 137, 139

Venus (VEE-nuhs) • a terrestrial planet, the second from the Sun in the Earth's solar system, 135-136, 138-139

voltmeter (VOHLT-mee-tuhr) • an instrument that measures the flow of electrons, 7

VY Canis Majoris (VEE WIY KAY-nuhs muh-JAW-ruhs) **(VY CMa)** (VEE WIY SEE EHM AY) • the biggest star we can see from Earth, 146

water (WAW-tuhr) • a molecule made of one oxygen atom and two hydrogen atoms, 17, 29-30, 95, 97

water vapor (WAW-tuhr VAY-puhr) • the gaseous state of water, 112, 115

weigh (WAY) • to measure how heavy something is, 6-7, 12, 39

Whale (WAYL) • a constellation in the Southern Hemisphere, 125, 128-129

yeast (YEEST) • a type of fungus, 57, 59, 60

Pronunciation Key

aa	past	(PAAST)
ay	date	(DAYT)
ah	far	(FAHR)
aw	saw	(SAW)
eh	pet	(PEHT)
ee	feed	(FEED)
ih	trip	(TRIHP)
iy	side	(SIYD)
oh	bone	(BOHN)
oi	boy	(BOI)
oo	rule	(ROOL)
ow	noun	(NOWN)
uh	fun	(FUHN)
uu	wood	(WUUD)
y	yet	(YEHT)

Image Credits

Cover and Title Page: By 1195798, Pixabay
Chapter 1: Page 1: By fstop123, iStock; Page 6. Photo by ThisisEngineering RAEng on Unsplash; Page 7. By Erwin Rossen, Public Domain
Chapter 2: Page 11. By R. Gino Santa Maria, AdobeStock; Page 12. Photo by National Cancer Institute on Unsplash; Page 13. By Robert Poorten, AdobeStock
Chapter 3: Page 16. By SpaceOak, AdobeStockAdobeStock; Page 20. Car battery, By warloka79, AdobeStock; Batteries, By New Africa, AdobeStock; Soda can, By Scanrail, AdobeStock; Grapes, By kovaleva_ka, AdobeStock; Citrus Fruit, By Africa Studio, AdobeStock; Page 21: Glass cleaner, By Aliaksei, AdobeStock; Banana, By xamtiw, AdobeStock; Dates, By olyina, AdobeStock; Soap, By pbombaert, AdobeStock; Page 22. Traffic light, Image by thorsteinn from Pixabay; Oven, By Yakobchuk Olena, AdobeStock; Thermometer, Image by NickiAnimations from Pixabay
Chapter 4: Page 25. Car battery, By warloka79, AdobeStock; Batteries, By New Africa, AdobeStock; Soda can, By Scanrail, AdobeStock; Grapes, By kovaleva_ka, AdobeStock; Citrus Fruit, By Africa Studio, AdobeStock; Glass cleaner, By Aliaksei, AdobeStock; Banana, By xamtiw, AdobeStock; Dates, By olyina, AdobeStock; Soap, By pbombaert, AdobeStock; Page 30. By Pixel-Shot, AdobeStock
Chapter 5: Page 32. Lemon, By baibaz, AdobeStock; Pickles, By nedim_b, AdobeStock; French fries, By sveta, AdobeStock; Pretzels, By ImagesMy, AdobeStock; Ice cream, By Olga Lyubkin, AdobeStock; Page 36. By Voyagerix, AdobeStock; Page 37. Image by congerdesign from Pixabay
Chapter 6: Page 42. Gulf of Alaska 2004 NOAA Office of Ocean Exploration
Chapter 7: Page 47. By Wire_man, AdobeStock; Page 48. Paramecium, By Wire_man, AdobeStock; Amoeba, By micro_photo, AdobeStock
Chapter 9: Page 57. Cheese, By Anatoly Repin, AdobeStock; Bread, By Noel, AdobeStock; Page 58. Top & Middle–Image by Andreas from Pixabay; Bottom– By Andreas, AdobeStock; Page 59. By Dmitry Knorre, AdobeStock
Chapter 10: Page 63. Scale, By Matthew, AdobeStock; Ruler, By The_Believer, AdobeStock; Timer, By Aleksandr, AdobeStock; Thermometer, Image by NickiAnimations from Pixabay; Page 64. Top–By JRJfin, AdobeStock; Bottom–By Dario Sabljak, AdobeStock; Page 65. Top–By David Lloyd, AdobeStock; Bottom–By Gorodenkoff, AdobeStock
Chapter 12: Page 74. By Augustas Cetkauskas, AdobeStock Page 77. By pavel1964, AdobeStock
Chapter 13: Page 81. By Ljupco Smokovski, AdobeStock; Page 84. By Volodymyr Shevchuk, AdobeStock
Chapter 14: Page 89. Top–By 2dmolier, AdobeStock; Bottom–By ChaoticDesignStudio, AdobeStock; Page 90. Top–By Delicious, AdobeStock; Bottom–By Vladislav Gajic, AdobeStock
Chapter 15: Page 96. NASA, Public Domain
Chapter 17: Page 110. By Pavla Zakova, AdobeStock; Page 111. By trekandphoto, AdobeStock; Page 112. By famiso, AdobeStock; Page 114. NASA, Public Domain; Page 115. By peopleimages.com, AdobeStock
Chapter 18: Page 120. Top–Courtesy of John Vicari; Bottom– Photo by Dorotheum, May 2011, Lot 40; Page 121. NASA, Public Domain
Chapter 19: Page 124. By Steve Jurvetson, CC BY SA 2.0
Chapter 20: Page 135-137 & 139. Planet images–NASA, Public Domain

www.ingramcontent.com/pod-product-compliance
Lightning Source LLC
Jackson TN
JSHW071144080525
83703JS00002B/9
9781941181805